PROJECT MANAGEMENT TOOLBOX

PROJECT MANAGEMENT TOOLBOX

Tools and Techniques for the Practicing Project Manager

Third Edition

Cynthia Snyder Dionisio

Russ J. Martinelli
Intel Corporation
USA

WILEY

Published by John Wiley & Sons, Inc., Hoboken, New Jersey.
Published simultaneously in Canada.

For general information on our other products and services or for technical support, please contact our Customer Care Department within the United States at (800) 762-2974, outside the United States at (317) 572-3993 or fax (317) 572-4002.

Wiley also publishes its books in a variety of electronic formats. Some content that appears in print may not be available in electronic formats. For more information about Wiley products, visit our web site at www.wiley.com.

Library of Congress Cataloging-in-Publication Data applied for:

Hardback ISBN: 9781394222063

Cover Design: Wiley
Cover Image: © Jorg Greuel/Getty Images

SKY10092489_120424

CONTENTS

PART V: PROJECT MONITORING, REPORTING AND CLOSURE TOOLS 245

14 SCHEDULE MANAGEMENT 247

17 PROJECT CLOSURE 315

PREFACE

I n the eight years since the second edition of the *Project Management Toolbox* was published, the practice of project management has changed significantly. Today, more than half of the projects use hybrid approaches to managing projects. Meaning that while predictive (waterfall) tools are still used, adaptive (Agile) tools and techniques are often integrated where applicable. The practice of separating the two ways of performing projects is waning and an attitude of "figure out the best way to deliver value" is taking over.

With that in mind, some of the tools that were predominantly used in the predictive projects of the past, such as a Time-Scaled Network Diagram or a Line of Balance Schedule, are not in the third edition. Several new tools have been introduced. The new tools can be used for predictive, adaptive, or hybrid projects, such as a Project Canvas, Project Roadmap, and Communication Matrix.

Some of the existing topics have been updated. The chapter on requirements has been refreshed. The intent of eliciting and managing requirements hasn't changed, but the tools used to do so have been updated. You will see a Requirements Management Plan and a Requirements Traceability Matrix rather than a Product Requirements Document and a Requirements Ambiguity Checklist.

There are also new chapters – Resource Planning, Advanced Risk Management, and Change Management. Some of these chapters have new content, and others reorganize content from the second edition.

The third edition of this book maintains the previous effective format of presenting a tool, describing how to develop it and then how to apply it, interspersed with examples and tips where applicable. Rather than having a separate section for the benefits of each tool, the benefits are indicated when introducing the tool.

It must be acknowledged that the world of project management is on the precipice of a huge change with the advent of artificial intelligence (AI). AI can be a tremendous asset when managing a project, but it cannot replace the fundamental understanding of project management and the knowledge of how to use tools to effectively manage projects. Therefore, while acknowledging the benefit of AI, this book does not describe how to use it in project management. After learning how to work with the tools in this Toolbox, you may choose to see how AI applies them, but first, you have to understand the fundamental tools, how to develop them and how to use them.

With thanks to the existing and future readers of this book. May it prove useful throughout your project management journey.

ACKNOWLEDGMENTS

Thank you to the many people who have helped in making this book a reality.

To the team at John Wiley & Sons, who continue to provide outstanding support and guidance. In particular, thanks go out to Margaret Cummins, Senior Editorial Director, and Kallie Schultea, Senior Editor. You are both wonderful women and outstanding professionals. It is an honor to work with you. Isabella Proietti and Jeevaghan Devapal have been helpful and professional in helping to get this book updated and published.

No one can take the time it takes to write a book without the understanding and forbearance of family. Thank you for your continued encouragement.

Finally, a huge thank you to project, program, and portfolio professionals everywhere. Every one of you is an inspiration. Keep uplifting our world!

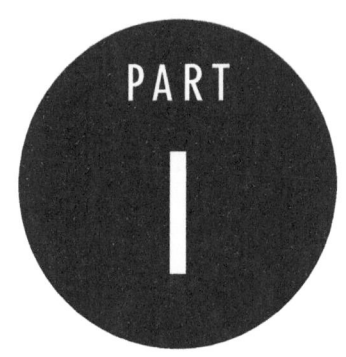

PART

I

The PM Toolbox

1

INTRODUCTION TO THE PM TOOLBOX

Project management tools support the practices, methods, and processes used to effectively manage a project. They enable the primary players on a project—the project manager, project team, executive leadership team, and governance body.

For purposes of this book, tools are considered to be processes, techniques, artifacts, software, or other job aids that assist in creating deliverables or project information. PM tools may be qualitative or quantitative in nature.

To illustrate, consider two examples, the Team Charter and a Monte Carlo Analysis. The Team Charter is an artifact that outlines how the team will work together on a project. A Monte Carlo analysis involves analyzing data that is generated from a software tool that uses an algorithm to quantify uncertainty around cost or schedule outcomes.

Note there is no mention of specific software tools here. While many PM tools discussed in this book exist in a software format, the focus is not on tool formats. Rather, the focus is on the use of tools to manage projects more effectively and efficiently.

A project management (PM) toolbox provides a set of tools that serves several purposes, such as:

1. Increasing the efficiency of the project players
2. Providing the right information to support problem solving
3. Providing relevant information for making decisions
4. Helping to establish and maintain alignment between business strategy, project strategy, and project outcomes.

The design of a PM Toolbox should align with the approach an organization takes for establishing project management methodologies and processes.

Project Management Toolbox: Tools and Techniques for the Practicing Project Manager,
Third Edition. Cynthia Snyder Dionisio, and Russ J. Martinelli.
© 2025 John Wiley & Sons, Inc. Published 2025 by John Wiley & Sons, Inc.

ALIGNING THE PM TOOLBOX

Organizations have a host of options when developing their methodologies and processes—they can be more standardized or more flexible. Generally, projects with a high degree of certainty do well with more standardization. Projects that face a high degree of uncertainty require more flexibility. The decision about how much to standardize project management methodologies and processes is driven by business strategy and by the types of projects needed to realize the business strategy.

The rationale behind standardization is to create a predictable process that prevents activities from differing substantially from project to project, and from project manager to project manager. Put simply, standardization saves project players the trouble of reinventing a new method and process for each individual project. As a result, the process is repeatable despite changes in customer expectations or management turnover.

The rationale behind flexibility is to give the project team the ability to explore, experiment, and iterate processes to reduce uncertainty. The players learn and adapt through multiple iterations in order to meet the needs of the project and the stakeholders.

When developing a PM toolbox, organizations should weigh the need for fixed and repeatable processes against the need for flexible and adaptable processes. This need may vary depending on the different departments or functions in an organization. For example, an engineering department may benefit from well-established policies, processes, and tools. In the same organization, the IT department may benefit from a more flexible and adaptable set of processes and tools. Both approaches are fine as long as they support the business strategy and are aligned with the project objectives.

Since the PM Toolbox is aligned with the PM methodology used, it is understandable that the level of standardization of the methodology impacts the standardization level of the PM Toolbox. For example, a methodology that is highly standardized will probably be supported by a highly standardized PM Toolbox.

CUSTOMIZING THE TOOLBOX

Developing a PM Toolbox is an evolutionary process. In a practical sense, PM toolboxes will look quite ad hoc at first. The tendency is to begin building the PM Toolbox with existing tools due to a project manager's familiarity with them. Thus, the early-stage PM Toolbox has more to do with familiarity of use than with standardization. As a firm begins to mature its project management practices, there is a greater understanding of the tools that are needed in the Toolbox.

Often, project managers assume that the PM Toolbox is of a one-size-fits-all nature. This is incorrect. The PM Toolbox reflects the project management methodology and types of projects the methodology serves.

Regardless of whether an organization's project management methods and processes are standardized, flexible, or semi-flexible, a PM Toolbox needs to be designed so that it aligns with both the PM methods and processes employed as well as the strategy of the project and the business strategies driving the need for the project. To accomplish this, a process for selecting and adapting the PM Toolbox is needed.

There are multiple options for customizing a PM Toolbox. Three of the most common are:

1. Customization by project size
2. Customization by project innovation
3. Customization by project type.

Each option has the purpose of showing which specific project management tools to select and adapt for the PM Toolbox. An in-depth knowledge of individual tools is a prerequisite to each of the options because you need to understand how each tool can support a project deliverable. This section describes the customization options and offers guidelines for selecting one of them for implementation.

Customization by Project Size

Some organizations use project size as the key variable when customizing a PM Toolbox. Their logic is that larger projects are more complex than smaller ones, or the size drives differences in project management methodology complexity. The reasoning here is that as the project size increases, so does the number of project management activities and resulting project deliverables associated with a project, and so does the number of interactions among them.

Since different project sizes require different processes and tools, we first need a way to classify projects by size and then customize their toolboxes. In Table 1.1 you can see examples of how different companies classify small, medium, and large projects.

Based on size, the companies determined the managerial complexity of the project classes and processes. The complexity influences the PM Toolbox make-up. A simplified example is shown in Table 1.2.

As Table 1.2 indicates, some of the tools in the toolboxes for projects of different size are the same, others are different. For example, all use the Lessons Learned (Chapter 17) because all projects need to learn from their performance. Since managerial complexity of the three project classes and their processes calls for different tools, some of the tools differ. For example, Earned Value Management (Chapter 15) is needed in large projects, but not medium or small projects.

Table 1.1: Examples of Project Classification by Size			
	Project Size		
Project and Company Type	Small	Medium	Large
Product development projects in a $1 Billion/year high-technology manufacturer	$1–2M	$2–5M	>$5M
Infrastructure technology projects in a $300M/year food processing company	<$50k	$50–150k	>$150k
Software development projects in a $40M/year customer relationship management software company	300–400 person-hours	1000–3000 person-hours	>3000 person-hours

Table 1.2: Examples of PM Toolbox Customization by Project Size				
Project Size	**Origination**	**Planning**	**Development**	**Closure**
Small	Project canvas	Scope statement	Summary status report	Project closure report
		WBS		
		Responsibility matrix		
		Milestone chart		
Medium	Project Charter	Stakeholder register Stakeholder analysis Communication matrix Scope statement	Summary status report	Project closure report
		WBS	Change management system	Post mortem review
		Responsibility matrix	Change log	
		Cost estimates	Critical path method	
		Critical path method	Cost baseline	
		Risk register Risk assessment Risk responses	Risk register	
Large	Project charter	Stakeholder register Stakeholder analysis Communication matrix Scope statement	Summary status report	Project closure report
	Complexity assessment	WBS		Post mortem review
		Responsibility matrix	Change management system	Project closure plan and checklist
		Cost estimates	Critical path method	
		Critical path method	Slip chart	
		Risk register Risk assessment Risk responses	Earned value management	
		Monte Carlo analysis	Risk register Monte Carlo analysis Risk dashboard	

When customizing the PM Toolbox by project size follow these steps:

■ Identify a small number of project categories
■ Define each category by the size parameter
■ Match the project size with the proper toolbox.

Note that while customization by project size offers advantages of simplicity, it also carries a risk of being generic, disregarding other situational variables. To some, these other variables may be of vital importance, as will be pointed out in the next section on customization by project family.

Customization by Project Innovation

The amount of innovation influences the tools in the PM Toolbox. For example, companies in the high-technology industry face an environment of dynamic technology change. Because of this, their portfolios have many quick time-to-market projects driven by the desire of their customers to continuously buy the latest and greatest technological products and services. Conversely, facilities management projects don't often have to contend with innovation and technology risk.

Innovation projects have more uncertainty. Uncertainty generally means more complexity, which requires more flexibility in the project management processes and the supporting toolbox. For example, as innovation grows:

- The more scope and requirements evolve
- The more the schedule becomes fluid
- Cost Estimates follow the fluidity of the schedules and scope.

A simple example reflecting these trends in adapting the toolbox for three levels of innovation could be Derivative Projects, Incremental Projects, and Breakthrough Projects.

- Derivative projects are those that have little to no innovation. The organization has done them before, the scope and requirements are well known, and there is little expectation of change.
- Incremental projects have some degree of innovation, often improving components or parts of a project. The outputs are mostly known, though there may be a need to iterate on a few aspects of the deliverables.
- Breakthrough projects deal with unknown, or evolving technologies. There may be uncertainty associated with requirements, technology, and solutions. These types of projects require a flexible approach to deal with the uncertainty and complexity associated with creating new solutions to problems or opportunities.

Table 1.3 shows an example of customizing the toolbox based on the degree of innovation.

As the table shows, the PM Toolbox for derivative and incremental projects are similar. However, breakthrough projects typically use a more Agile or adaptive approach that allows for evolving and changing scope.

Customization by Project Type

While the previous two approaches to PM Toolbox customization rely on one dimension each—project complexity and project innovation—customization by the project type uses two dimensions.[1]

Table 1.3: Customizing the Toolbox by Project Innovation				
Project Innovation	Origination	Planning	Development	Closure
Derivative projects	Project charter	Milestone chart	Summary status report	Project closure report
	Scoring models	Requirements specification		
		WBS		
Incremental projects	Project charter	Stakeholder analysis Scope statement	Summary status report	Project closure report
	Scoring models	WBS or PWBS	Change log	Change log
		Requirements specification	Critical path method	Lessons learned
		Cost estimates	Budget consumption chart	
		Critical path method	Risk register	
		Risk register Risks analysis Risk responses		
Breakthrough projects	Project roadmap	Stakeholder engagement plan	Backlog	Project closure report
	Complexity assessment	Requirements elicitation plan	Requirements traceability matrix	Post mortem review
		Backlog	Task board	Lessons learned
		Release planning	Release planning	
			Sprint retrospective meetings	

This model shows a grid with each dimension. Each dimension includes two levels: (1) innovation of the capability under development (low, high) and (2) project complexity (low, high). This helps to create a two-by-two matrix that features four types of projects—routine, administrative, technical, unique (see Figure 1.1).

A routine project is one having a low level of capability innovation (less than half of the technologies are new) and low complexity (few cross-project interdependencies). Due to the low levels of innovation and complexity, the project scope can normally be frozen before development begins or early in the development phase. Scope also remains fairly stable with few changes. With scope remaining stable, project scheduling, cost management, and performance management are also quite static.

Typically, routine projects are performed within a single organization or organizational function (for example, infrastructure technology). Examples include the following:

- Continuous improvement project in a department
- Upgrading an existing product
- Developing an updated model in a product line
- Expanding an established manufacturing line.

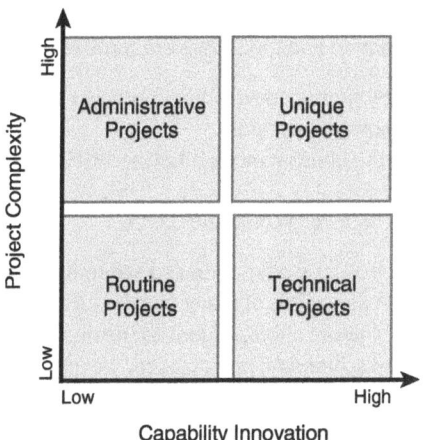

Figure 1.1: Four Project Types

Administrative projects are similar to routine projects in terms of innovation. Business goals and scope are normally well defined, stable, and detailed. The added complexity requires the coordination of multiple organizational functions and the mapping of the many functional interdependencies, but the lack of capability innovation allows for standard scheduling techniques. The same added complexity generally means larger project size, with higher financial exposure, justifying the need for detailed bottom-up cost estimates reconciled with financial targets contained in the project business case. Risk is primarily related to the increased number of interactions between the functions and project team; therefore, additional risk planning and analysis is required.

Some examples of administrative projects are as follows:

- Corporate-wide organizational restructuring
- Deploying a standard information system for a geographically dispersed organization
- Building a traditional manufacturing plant
- Upgrading an enterprise computer system.

Technical projects consist of more than 50% of new technologies or features at the time of project origination. This creates a higher degree of uncertainty that requires project flexibility. The goals, scope, and WBS are simple due to the low level of complexity, but they may take longer to fully define. The rolling wave or similar approach can be used, meaning that only the schedule for the following 60–90 days can be planned in detail, while the remainder of the program schedule is represented only by milestones. Similarly, cost estimates are fluid as well. A detailed cost estimate for the next 60–90 days can be developed, while cost estimates for the remainder of the project are at the summary or rough order of magnitude level. The increased technical innovation results in increased technical risk and the need for a more rigorous risk management implementation and tools. Here are some examples:

- Reengineering a new product development process in an organization
- Developing a new software program
- Adding a line with the latest manufacturing technology to a semiconductor fabrication plant
- Developing a new model of a computer game.

For unique projects, business goals, detailed scope definition, and WBS development take time to evolve as a result of many new features and cross-project interdependencies. The evolving nature of scope leads to the need for fluid schedules. Project mapping and rolling wave scheduling processes can be used to contend with the fluidity. Similarly, cost estimates for milestones are more detailed in the near term and more summary level for the longer term. A high degree of project complexity exists due to multiple organizational functions required to execute unique projects, requiring integration tools. Combined capability innovation and project complexity push risks to the extreme, making it the single most challenging element to manage. In response, a rigorous risk management plan is needed, as well as a combination of tools such as a Monte Carlo Analysis and a Risk Dashboard (Chapter 11). Example technology projects include:

- Building a new light rail train system for a city
- Developing a new generation integrated circuit
- Developing a new software suite.

Now that we've defined the four project types, we can move on to the next step—describe how the two dimensions impact the construction of the PM Toolbox. Taken overall, the growing technical innovation in a project generates more uncertainty, which consequently requires more flexibility in the tools chosen. Figure 1.2 shows examples of several project management tools that are adapted to account for different processes driven by different project type.

A summary comparison of the tools for the four project types reveals that they use similar types of tools. For example, all use the WBS. Still, when the same type of tool is used, there are differences in their structure and how they are used. Consider, for instance, Gantt and Milestone Charts. Both are used in the routine and unique projects, but terms of use are significantly different. This is the situational approach—as the nature of the project management processes change, so does the PM Toolbox.

Which Customization Option to Choose?

The three options for customizing the PM Toolbox each has its advantages, disadvantages, and risks. Each option fits some situations better than others. Table 1.4 provides some insight on how to choose the option best suited for your needs. Customization by project size is a good option when an organization has projects of varying size and

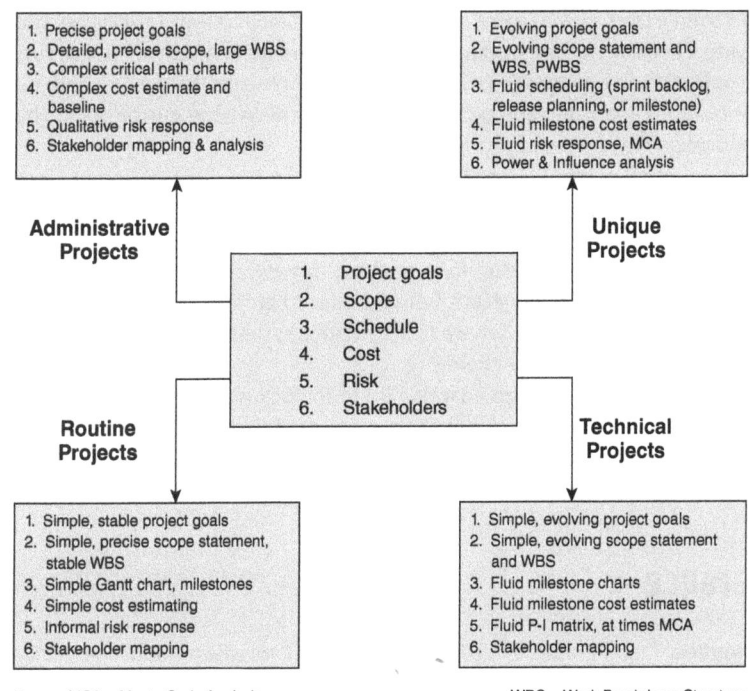

Figure 1.2: Customizing the PM Toolbox by Project Type

Table 1.4: Project Situations and PM Toolbox Customization

Situation	Customization by Project Size	Customization by Project Innovation	Customization by Project Type
Simplest start to PM Toolbox customization	✓		
Projects of varying size with mature capabilities	✓		
Projects with both mature and innovation aspects, size not an issue		✓	
Projects with strong industry or professional culture		✓	
Projects of varying size with both mature and innovative capabilities			✓
Need for a unifying framework for all organizational projects			✓

needs a simple start toward more mature forms of customization. In addition, projects of varying size characterized by mature processes lend themselves well to this customization option. In an organization that has a stream of projects that feature both mature and innovation capabilities and project size is not the main issue, customization by project innovation may be the best option.

Customization of the PM Toolbox by project type is a good option in situations where an organization has a lot of projects that significantly vary in size and in innovation of the solutions, such as a portfolio of government research and procurement projects. Organizations searching for a unifying framework that can provide the customization for all types of projects—from facilities to product development to manufacturing process to customer service to information systems—may find customization by project type an appropriate choice.

Effectively constructing and adapting a PM Toolbox is predicated upon the user's knowledge of individual project management tools. To help increase your knowledge of individual tools, the remaining chapters detail a multitude of useful tools that can be chosen for inclusion in your own PM Toolbox.

Reference

1. Boutros, T. and Purdie, T. (2013). *The Process Improvement Handbook: A Blueprint for Managing Change and Increasing Organizational Performance*. New York, NY: McGraw-Hill.

PART

II

Project Start Up Tools

2

PROJECT SELECTION

This chapter is intended to help with the selection of projects based upon highest value to an organization. We describe a number of tools that can be added to a PM Toolbox which help to evaluate the value a project offers, the benefits it can deliver, and how well it aligns to business strategy. There are several tools that can be used to initiate a project so it is set up to run efficiently.

Most organizations have many more products, services, infrastructure solutions, or transformational change ideas than available resources to execute them. As a result, an organization must find a way to prioritize demands for its limited resources. Project selection techniques are used to identify and prioritize projects that best support attaining the enterprise business goals. Projects are ranked and prioritized based on a set of criteria that represent business value to the organization. Value in this context means something important or of worth. Senior management can then allocate available resources to the highest value and most strategically significant projects. This is sometimes referred to as resource capacity planning.

The outcome of projects must contribute to the long-term viability of the enterprise by delivering both short-term and long-term value. Project selection tools are designed to help with the selection of projects that will maximize the value of the investment made in a portfolio of projects. The mix of projects within a portfolio are in a state of constant flux, as an organization reevaluates its selections on an ongoing basis to respond to changes in the business environment and other needs. Project selection tools allow an organization to select projects for initiation and termination as conditions change.

There are of course many tools for selecting projects. This chapter presents the tools that can be used across a variety of industries and organizations.

BENEFITS MAP

Within the practice of project management, value is often associated with the generation and delivery of benefits. In turn, value management is often described in terms of

Project Management Toolbox: Tools and Techniques for the Practicing Project Manager,
Third Edition. Cynthia Snyder Dionisio, and Russ J. Martinelli.
© 2025 John Wiley & Sons, Inc. Published 2025 by John Wiley & Sons, Inc.

benefits management. Therefore, project managers must be able to describe the value proposition of their projects in terms of the business benefits the project will create for the organization. The management team then evaluates the value proposition of the project opportunities to prevent an over commitment of its limited resources.

For many organizations, the objective measure of value is in financial terms. Other organizations make a qualitative measure of value by assessing the business benefits delivered in alignment to strategic goals. In either approach, benefits are the outcomes of a project that provide value to the organization in return for the investment made in the project.

The Benefits Map also provides a systematic process to assess business benefits as part of the project's cost/benefit analysis, which is a critical element of the business case of a project and in project selection.

We begin by looking at the qualitative assessment of benefits and value.

Enabling Benefits Management

Benefits management is about managing the achievement of the intended business results. A useful tool for benefits management is the development of a benefits map. Creating a benefits map involves charting the path from the organization's business goal to the success factors, which are then quantified with objectives. The objectives are then connected to deliverables.

The Benefits Map can be depicted in visual form to provide a useful means to demonstrate alignment of project deliverables to the business success factors for a project, to the expected business goal as displayed in Figure 2.1.

This is critical information to first establish the overall vision and scope for a project, then communicate how the project contributes to the objectives, and finally track the execution of the project to final delivery of business goals.

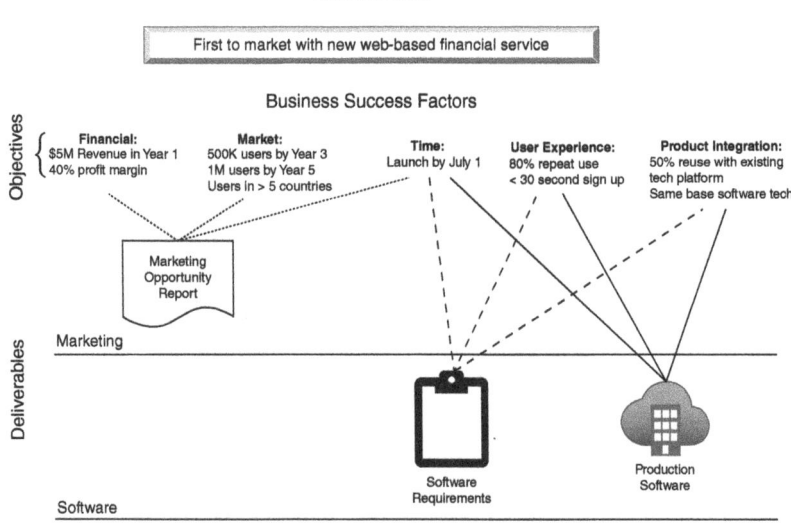

Figure 2.1: Sample Project Benefits Map

This tool is used to assist in the characterization of how specific project objectives are met. However, benefits maps can become complex and confusing due to the one-to-many relationships between project deliverables to the objectives. The critical component in building an effective Benefits Map is to ensure each project deliverable is mapped to an objective, and every objective to the business success factors.

Developing a Benefits Map

The development of a Benefits Map is not a simple exercise because it requires knowledge of both business strategy and project execution. The major steps in creating a benefits map are described below.

Identify the Strategic Business Goals

The first activity involved in developing a Benefits Map is to define the strategic business goals that are underwriting the need for the project. Strategic business goals define what the company wants to achieve within a defined period of time and are normally defined at two levels of an organization: corporate strategic goals and business or operating unit strategic goals.[1]

The purpose of corporate-level strategic goals is to align the various business units toward a common purpose and direction, while business unit strategic goals serve to focus the functional departments and work efforts of the people within a business unit.[2] Every enterprise is unique and, therefore, every enterprise will have its own set of strategic goals. It is common, however, to find strategic goals centered in a number of areas including:[3]

- Profitability
- Competitive Position
- Employee Relations
- Product or Solution Leadership
- Productivity
- Employee Development
- Public Responsibility.

Businesses do not normally identify strategic goals in all the areas presented above, but rather only in those that align with and support attainment of the corporate mission.

When developing a benefits map, it is important to identify the strategic goal or goals that the project is intended to help achieve. This is accomplished through knowledge of the business aspects of an enterprise and through a series of discussions with the senior leaders of the organization.

Define the Business Success Factors

With the strategic business goals identified, it is time to define how the achievement of the strategic business goals will be measured. The business success factors transform the business results derived from strategic goals into a set of categories that guides the planning and delivery of project work. There should be three to six business success factors. During the early stages of a project, the factors are used to align the program sponsor, executive stakeholders, and the project team on what project success looks like.

Business success factors are then quantified into specific objectives. The metrics used to define successful achievement of the business success factors for the project are the objectives. The objectives are used by the project team to establish the end state they are working toward.

Identify Project Deliverables

You can use a preliminary version of a work breakdown structure (Chapter 6) to identify the primary deliverables for the project. These are normally defined at level two or level three of the breakdown structure.

Once project deliverables are identified, it normally works well to categorize the deliverables by ownership. In other words, group all deliverables by each project sub-team that will be responsible for creating and delivering the outcome.

As shown in Figure 2.1, you can document the deliverables in swim lanes. Even though the swim lanes look linear in nature, there is no time element to the benefits map, so the outcomes for each team do not have to be documented in time sequence order. In fact, it is better to try to align them with the business success factors they support.

Perform the Mapping

This step simply involves using interconnecting lines to graphically illustrate the relationship between project deliverables, project objectives, and the business success factors. The use of color-coded lines associated with the particular project outcomes is especially helpful for larger projects.

Validating the Results

The final step in creating a Benefits Map is to validate that each project deliverable directly supports the achievement of a project objective, and that the project objective directly supports the achievement of the business success factor it is associated with.

When completed a project manager has a visual mapping of how a project deliverable supports the strategic business goals of the firm via a direct link between project objectives and business success factors.

Using a Benefits Map

Along with the work breakdown structure, the Benefits Map is a useful tool for establishing the overall scope of a project. The Benefits Map can also be used to communicate to top management, the project team, and other stakeholders how the strategy of the organization and project are melded together, and how each benefit will be realized.

The Benefits Map is intended to be used throughout the life of the project to analyze consequences caused by adjustments and changes as they occur to the original project strategy and scope. The first use of the benefits map normally occurs as part of the business case development process where a high-level mapping of benefits to project objectives to strategic intent is established. Further detail is then added during detailed planning when the full comprehension of scope and traceability of project deliverables to business benefits is necessary. During the project it enables focused tracking and monitoring of progress toward realization of the benefits as part of the project reporting process and establishes an effective means for evaluating success of a project from a benefits realization perspective.

Variations

At times, business benefits are described in terms of solutions to problems that have been identified. If this is the case, a variation of the Benefits Map, often called the Objectives Tree, is better suited to assess project value. The goal defined in the Objectives Tree can be used to refine the project objectives by outlining the high-level requirements.

An Objectives Tree provides a visual representation that allows you to quickly and completely articulate the scope of the problem you are attacking. It is used primarily in the earliest stages of project definition.

As illustrated in Figure 2.2 the objectives—which are the means to solving a problem—are decomposed into high-level requirements.

When using the Objectives Tree to solve a problem, the core problem is reworded into an objective that describes a desired end state. For example, suppose we are the new product development team in charge of defining, designing, and producing our company's tablet products. A primary objective we would likely have is to specify the next tablet that will be designed and manufactured. This primary objective would then be decomposed into more detailed objectives that define solutions to the primary objective. Decompose at least three levels and then work your way back up the tree to validate that the high-level requirements are sufficient to achieve the objectives at the next highest level. The resulting Objectives Tree can be used as the basis for defining more detailed requirements.

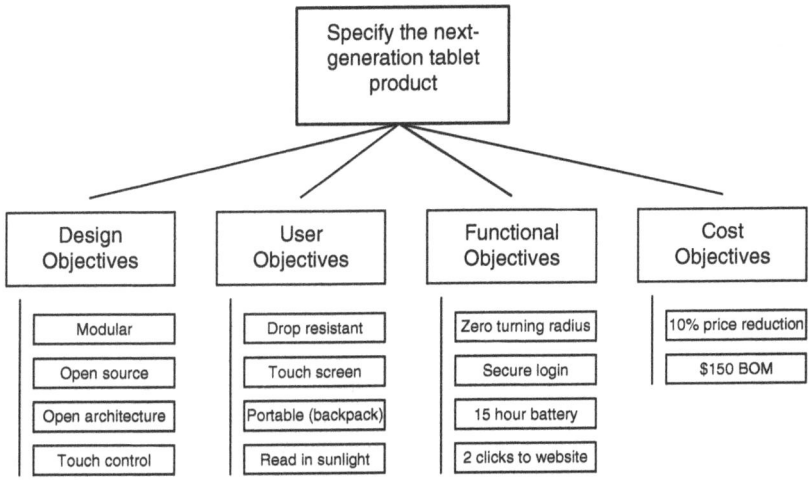

Figure 2.2: Example Objectives Tree

ECONOMIC METHODS

When quantitative methods are used to define project value to guide the project selection process, they are normally economically focused. Because quantitative measures are readily understandable, they enable the decision maker and project managers to communicate more readily about financial considerations of projects. They also make it easier to compare projects with other opportunities that are vying for capital investment.

Once the necessary data are obtained, they are easy to compute. They also make sensitivity testing easy by adjusting discount rates, time, and other input data. Alternate scenarios about future costs and returns can be compared to test various assumptions being made about project value against future uncertainty. Here we discuss the three most common economic methods for describing project value: Payback Time, Net Present Value (NPV), and Internal Rate of Return (IRR).

Payback Period

Payback period is simply the length of time from when the project is officially initiated until the cumulative cash flow (or cost savings) becomes positive. At that point all the funds invested in the project have been recovered. As you can see in the three project cash flow scenarios in Table 2.1, the cash flow for Project 1 turns positive in 7 years, for Project 2 in 5 years, and for Project 3 in 8 years.

Payback period is a very conservative criterion and provides more protection against future uncertainties than do either of the other economic methods (NPV and IRR). However, there are some weaknesses:

- It is insensitive to project size, since a project with massive investment requirements may still have short payback time
- It doesn't account for the time value of money
- It takes no account of future economic potential once payback is achieved.

Net Present Value

Net present value accounts for the time value of money, in that a dollar a year from now is worth less than a dollar today due to inflation. NPV discounts both future costs and revenues by the interest rate, according to the formula:

$$NPV(i,N) = \sum_{t=0}^{N} \frac{R_t}{(1+i)^t}$$

In this formula, R_t represent the net cash flows (cash inflow–cash outflow, at time t), i is the discount rate or the rate the company pays for borrowed money, expressed as a decimal fraction, and N is the total number of periods (years, months, etc.).

Spreadsheet programs can be used to calculate NPV directly. You only need to enter the discount rate and the values into the NPV function. The result is computed and displayed in the cell holding the function. Using the NPV function, the NPV for three projects at a discount rate of 5%, 10%, and 15% is shown in Table 2.2.

Table 2.1: Project Payback Period Examples								
	Project 1			Project 2			Project 3		
Year	Cost	Revenue	Cum Cash	Cost	Revenue	Cum Cash	Cost	Revenue	Cum Cash
1	20		−20	30		−30	20		−20
2	50		−70	80		−110	50		−70
3	80		−150	105	147	−68	78		−148
4	120		−270	110	154	−24	80		−228
5	200	200	−270	125	175	26	150		−378
6	450	630	−90	150	210	86	375	525	−228
7	500	775	185	160	220	146	525	735	−18
8	500	775	460	170	223	199	600	840	222
9	450	700	710	175	230	254	800	1120	542
10	450	650	910	170	228	312	750	1200	992

Table 2.2: Project Net Present Value Examples		
Project 1	Project 2	Project 3
NPV at	NPV at	NPV at
5% discount rate = $5,283	5% discount rate = $2,320	5% discount rate = $6,400
10% discount rate = $2,841	10% discount rate = $1,254	10% discount rate = $3,275
15% discount rate = $1,563	15% discount rate = $688	15% discount rate = $1,679

You can see that the more the future is discounted (i.e., higher discount rate), the less the NPV of the project. When comparing the three projects for project selection, Project 3 delivers greater value than either Project 1 or 2 at all discount rates. That is, the higher the NPV, the greater the economic value of a project.

NPV accounts for the magnitude of the project and the discount rate. However, it incorporates estimated future revenues that may not actually materialize.

Internal Rate of Return

Internal rate of return (IRR) is simply the discount rate at which NPV for the cash flow is zero. There is no closed-form formula for it. IRR must be computed iteratively by "honing in" on the exact discount rate that produces an NPV of zero. Most spreadsheets have an IRR function that allows the user to obtain an IRR. You need to only enter the list of values and a guess value of IRR. The function then carries out the iterative calculation. For the projects shown in Table 2.2, the IRR values are as follows:

Project 1 IRR = 42%

Project 2 IRR = 40%

Project 3 IRR = 36%

By the IRR criterion, Project 1 is superior in economic value to the other two projects. While IRR discounts future values, it takes no account of the size of the project. Project 3 promises a significantly greater total return than Project 1, but those returns are farther in the future and follow a longer course of investment before cash flow turns positive. Hence, the IRR for Project 3 is lower than for Project 1.

Using Economic Methods

Financially driven quantitative measures of value require data about revenues (or cost savings) and costs expected to result from a project. They are thus appropriate primarily for capital projects and projects intended to improve existing capabilities (products, services, or infrastructure). In such cases they allow a direct comparison of such projects with alternative capital investments.

You should ensure that the same economic method is used for all project valuation calculations to provide a direct comparison of projects. Also ensure you know your organization's standard discount rate before performing the calculations. For a comparison of the three economic methods, see the example titled "Choosing an Economic Method."

Choosing an Economic Method

As demonstrated, the three methods can give differing results on the same set of projects. The method chosen depends on the considerations important to the decision maker.

NPV is best used for projects with large payoffs, though it provides little protection against future uncertainties. IRR likewise gives little protection against future uncertainties and may tend to give preference to a project with modest total payoff but high return on a modest investment. Payback period provides more protection against future uncertainty, but it does not factor in the size of project payoff, nor of the discounted value of future costs and revenues.

The consideration that is most important will govern the choice of methods. Moreover, since these methods treat projects as capital investments, it would be appropriate to use whichever method the company uses for evaluating its other capital investments, thus allowing a direct comparison between project investments and other investments.

SCORING MODELS

A Scoring Model is a list of criteria the decision maker wishes to take into account when selecting projects from the list of candidate projects. Projects are then rated by decision makers on each criterion, typically on a numerical scale. Multiplying these scores by weights and aggregating them across all criteria will produce a score that represents the merit of the project. Higher scores designate projects of higher value.

The scoring model is that it can be tailored to fit the decision situation, taking into account objective and subjective goals and criteria deemed important for the decision. This prevents putting a heavy emphasis on financial criteria that tend not to be reliable early in the project life. With such an approach, decision makers are forced to scrutinize each project on the same set of criteria, focusing rigorously on critical issues but recognizing that some criteria are more important than others (by means of weights).

Developing a Scoring Model

The purpose of a scoring model is to help maximize the value of the selected projects for the company; therefore, understanding which of the company's strategic goals a project supports is a key point. Once the strategic goals are understood, there are three steps to develop a Scoring Model:

1. Define the scoring criteria
2. Rate the value of each criteria
3. Identify measures for the criteria

Define Scoring Criteria

The first step in developing a Scoring Model is identifying an appropriate set of scoring criteria that reflects the strategic, financial, technical, and behavioral situation of the company. To be used effectively, the criteria need to be described in specific and measurable terms. It is a best practice to have no more than five scoring criteria.

Scoring criteria depends on the type of project and the purpose of the project. For a new R&D project you may choose criteria such as:

- Market size
- Strategic alignment
- Competitive landscape
- Financial metrics.

Process improvement projects would have different criteria as would new product development projects.

Constructing the Model

To construct a Scoring Model, you must understand and resolve several issues:

1. The form of model you want to use
2. Categories of scoring criteria you want to use
3. Value and importance of the criteria
4. Measurement of the criteria.

First, we will deal with formation of the model. A *generic* scoring model would have the following form:

$$Score = \frac{A(bB + cC + dD)(1 + eE)}{fF(1 + gG)}$$

The symbols *A*, *B*, *C*, *D*, *E*, *F*, and *G* represent the criteria to be included in the score for the project. The value of each criterion for a given project is substituted in the formula. The symbols *b*, *c*, *d*, *e*, *f*, and *g* represent the weights assigned to each criterion. In the model, the criteria in the numerator are benefits, while the criteria in the denominator are costs or other disbenefits. The values of the criteria are project-specific and are normally provided by the project team.

This model uses three categories of criteria:

1. Overriding criteria (e.g., *A*). These are factors of such great importance that if they go to zero, the entire score should be zero. For instance, factors to be included in the model might be measures of performance such as efficiency or total output. A performance measure of zero should disqualify a project completely, regardless of any other merit.

2. Tradeable criteria (*B*, *C*, *D*, *F*). These are factors that can be traded against one another; a decrease in one is acceptable if accompanied by a sufficient increase in another. For instance, a designer may be willing to trade between reliability and maintainability, so long as "cost of ownership" remains constant. In this case, the weights would reflect the relative costs of increasing reliability and making maintenance easier. Cost *F* is shown as a single criterion that is relevant to all projects. Typically, this would include monetary costs of the project. This might be disaggregated into cost categories such as wages, materials, facilities, and shipping, if there is the possibility of a trade-off among these cost categories. If no such trade-off exists, the costs should simply be summed and treated as a single factor.

3. Optional criteria. These are factors that may not be relevant to all projects; if they are present, they should affect the score, but if they are absent, they should not affect the score. Note that either costs or benefits may involve optional factors. For instance, *E* in the formula represents a benefit that may not be a consideration with all projects. It should be counted in the score only if it is relevant to a project. For example, this might be a rating of ease of consumer use, which would not be relevant to a project aimed at an industrial purpose. *G* in the formula represents an "optional" cost than might not be relevant to all projects. Typically, this type of cost is one in which the availability of some resource is a more important consideration than its monetary costs. For instance, there may be limits on the availability of a testing device, specialized computers, or a scarce skill such as a programmer. In such a case, the hours or other measures of use should be included separately from monetary cost and should apply only to those projects requiring that resource.

The second issue focuses on value and importance of criteria. Once the form of the model is selected, the designers of the model need to distinguish between the value of a criterion and the weight or importance of than criterion. In the preceding formula, *B*, *C*, and *D* are the values of their respective factors for a specific project, while *b*, *c*, and *d* are the weights assigned to those factors, reflecting the importance assigned to them by the decision maker. In the case of the tradable factors, the ratio *b/c* represents the trade-off relationship between factors *B* and *C*. If B is decreased by one unit, C must be increased by at least the amount *b/c* for the sum of the tradable factors to remain constant or increase. That is, the decision maker is willing to trade one factor for another according to the ratios of their weights, so long as the total sum remains constant or increases.

Finally, the third issue to resolve is the measurement of criteria. Some criteria are objectively measurable, such as costs and revenues. Others, such as probability of success or strategic importance, must be obtained judgmentally. Scoring models can readily include both objective and subjective or judgmental criteria. It is helpful if the judgmental criteria are estimated with a scale with descriptor phrases, to obtain consistency in estimating the magnitude of the factor for each project. The estimates should be made on some convenient scale, such as 1–10. For an example see "An Example of Subjective Measure of Criteria." A similar scale should be devised to aid in making estimates of each of the criteria to be obtained judgmentally, as was done in the example in Table 2.3.

An Example of Subjective Measure of Criteria

10 All skills are in ample supply
9 All skills are available with no excess
8 All technical skills are available
7 Most professional skills are available
6 Some technical skill retraining is required
5 Some professional skill retraining is required
4 Extensive technical skill retraining is required
3 Extensive professional skill retraining is required
2 Most technical skills must be hired
1 Most technical and professional skills must be hired
0 All technical and professional skills must be hired.

Most scoring models are more complex than a simple sum of criteria. Suppose the factors we wish to include in the score are probability of success, payoff, and cost. Suppose further that we are willing to trade payoff and probability of success (e.g., we are willing to accept a project with higher risk if the payoff is high enough), and we think payoff is twice as important as either probability of success or cost. Probability of success and payoff are benefits, where cost is a cost or disbenefit. Then the scoring model will be as follows:

$$Score = \frac{P_{Success} + 2 \times Payoff}{Cost}$$

The designer of the scoring model is free to include whatever factors are considered important, and to assign weights or coefficients to reflect relative importance.

Table 2.3: Rating a New Product Development Project with a Scoring Model			
Criteria/Factors (Scored 0–10)			
Criteria/Factor	Item	Score out of 10 (points)	Average Criterion/Factor Score (Points)
Strategic positioning	Degree of project's alignment with business unit strategy (Strategic significance)	8	8.0
Product/Competitive advantage	Unique product functionalities	8	8.0
	Provides better customer benefits	9	
	Meets customer value measures better	7	
Market appeal	Market size	8	7.0
	Market share	8	
	Market growth	6	
	Degree of competition	6	
Alignment with core competencies	Market alignment	8	7.0
	Technological alignment	7	
	Manufacturing alignment	6	
Technical merit	Technical gap	9	8.0
	Technical complexity	6	
	Technical probability of success	9	
Financial merit	Expected net present value	9	8.0
	Expected internal rate of return	9	
	Payback time	7	
Total project score			130 out of a possible 170 points (77%)

Scoring Projects

When the criteria for the model have been selected, the form of the scoring model chosen, weights established, and measurement scales defined, you are ready to rank the candidate projects. Note that while the decision maker(s) must obtain the criteria and their weights from management, this is a one-time activity. The project-specific data for individual projects will in most cases come from those proposing the project. They will provide either objective data (e.g., costs, staff hours, machine use) or ratings based on the scales the decision maker has established. In some cases, project-specific data may be obtained from sources other than the project originators. For instance, data on probability of market success or payoff might be obtained from marketing rather than from R&D. While the data must be obtained for each project being ranked, the criteria and their weights will remain fixed until management decides they must be revised.

In most cases, the project data will be in units that vary in magnitude: probabilities to the right of the decimal, monetary costs to the left of the decimal, scale rankings in integers, and so forth. It is necessary to convert all the factors to a common range of values. Assuming that the project-specific values are approximately normally distributed, the result should be standardized values ranging from about −3 to about +3. These must now be restored to positive values. If any of the original values in a column was zero, the standardized value for that factor should also be zero. Add to every value the absolute value of the most negative number in the column resulting from the subtraction and division process. This will result in standardized values ranging from zero to approximately 6. If none of the original values was zero, add to every number 1 plus the absolute value of the most negative number. This will give values ranging from 1 to approximately 7. These standardized values should then be substituted in the model. Each project then receives a score based on weights supplied by management, and project data supplied by the project originators.

Using a Scoring Model

Table 2.4 shows the results of the model applied to standardized scores. The rows have been re-ordered in decreasing magnitude of the score. If the standardized values are available in a spreadsheet, the process of computing project scores is trivial. Likewise, sorting the projects in order of scores is readily accomplished using a spreadsheet. In this example, Project 4 is the highest ranked project. The other projects fall in order of their score. The next step would be to approve projects starting from the top of the list and working down, until the budget and/or resources are exhausted. Note that the difference between projects 8 and 1 comes only in the third significant figure. Since most of the original data were good only to one or two significant figures, this difference should not be taken seriously.

While scoring models can be used for any type of project, they are especially useful in the earlier phases of a project life cycle when major project selection decisions are made. Take, for example, new product development projects. In the earlier project phases, market payoff is distant or even inappropriate as a measure of merit. In such projects, considerations such as technical merit—a frequent criterion in scoring models—may be of greater significance than economic payoff. Selection of other types of projects, large and small, widely relies on scoring models as well. The final score is typically used for two purposes:

1. *Go/kill decisions*. These are located at certain points within the project management process, often at the end of project phases. Their purpose is to decide which new projects to initiate and which of the existing ones to continue or terminate.
2. *Project prioritization*. This is where resources are allocated to the new projects with a "go" decision, and the total list of new and existing projects, which already have resources assigned, are prioritized.

Table 2.4: Ranking of Projects Using the Scoring Model				
Project	Cost ($k)	P (S×s)	Payoff ($M)	Score
4	1.89	2.67	3.35	4.96
6	2.12	3.38	2.78	4.22
3	1.00	2.13	1.00	4.13
5	2.17	3.33	2.78	4.10
8	2.51	3.88	2.37	3.43
1	1.45	2.13	1.42	3.42
12	3.58	3.56	4.22	3.35
11	3.70	3.61	3.34	2.78
2	1.62	1.00	1.58	2.57
7	2.49	3.67	1.34	2.56
14	4.44	3.78	3.54	2.45
16	6.39	3.88	5.74	2.4
9	3.11	3.56	1.91	2.38
10	4.13	3.65	1.91	2.38
13	4.11	3.98	2.53	2.20
15	5.43	3.86	2.88	1.77

The principles behind the scoring models are relatively simple. They trim down the complex selection decision to a handy number of specific questions and yield a single score, which is a helpful input into a project selection effort. Several studies showed that they yield good decisions. For example, Procter and Gamble claims that their scoring models provide an 85% predictive ability.[4]

VOTING MODELS

For some organizations, numeric scoring models can become complex and cumbersome. We have found that when scoring models fail within an organization, they do so for a couple of primary reasons. First, teams get embroiled in trying to design the perfect scoring model. An enormous amount of time and energy can be lost in debating the right scoring criteria, the definition of numeric value for each criterion (e.g., how to describe what a scoring value of 1 is, versus 2, versus 3, etc.), and which criteria are more important than the others so weights can be assigned.

Second, numeric scoring models quite often fail to provide good scoring separation between projects. One executive from a financial services organization describes this outcome as "the scores tend to munge to the middle" to form a bell curve effect. When this happens, it gets very difficult to evaluate which projects are the best investment choices.

For these reasons, some organizations adopt an approach that provides the necessary structure and information to make project investment choices but relies on the experience and judgment of a cross-section of informed stakeholders. Voting models are effective in facilitating this judgmental approach by providing a technique to tap the collective knowledge of a group of experts with diverse perspectives. Voting Models take advantage of these diverse perspectives to create a clear understanding of the highest priority projects, increasing the organizational knowledge about the value proposition of each project, creating broad buy-in of project priorities across the organization, and clearly separating the wheat from the chaff.

Developing a Voting Model

The process for developing a Voting Model is very similar in nature to developing a numeric Scoring Model, but simplified. The primary difference is in how project value is evaluated and scored. Simplification is achieved through the limitations imposed on the scoring structure. Project value for any one criterion is limited to three values (1–2–3, H–M–L, etc.), and no weighting of criteria is needed. The major steps in developing a Voting Model are described below.

Identify the Stakeholders

Critical to a successful outcome in using Voting Models is the assembly of the right set of stakeholders who will have a vote on project value and prioritization. The intent is to assemble a set of stakeholders who have a vested interest in the prioritization outcome and represent a good cross-section of functional perspectives. It is advisable that no more than twelve to fifteen stakeholders participate in a voting event.

Develop Value Propositions for Each Project

For each of the candidate projects, a brief value proposition should be prepared for use in the voting event. Limit the amount of content to that which can be presented to the stakeholders within three to five minutes at maximum.

Create the Prioritization Criteria and Value Anchors

For voting models, the prioritization criteria have to be limited to the critical few criteria. The maximum number of criteria should be limited to no more than five. For each criterion, a description of the criterion and the value anchors are prepared prior to the voting event. Table 2.5 shows an example of criteria and value descriptions.

Create a Voting Template

Voting Models are best used in a facilitated work session where all critical stakeholders can be assembled and provided the opportunity to discuss and debate the value of

Table 2.5: Example Criteria Description and Value Anchors

Criteria	A	B	C
Monetary Value	Clear path to money, high ROI	Clear path to money, low ROI	No clear path to money
	> $100M ROI (3 yr)	$50–100M ROI (3 yr)	<$50M ROI
Strategic Value	Severe competitive threat	Moderate competitive threat	Low competitive threat
	Must do for leadership or time critical neutralizer	Moderate for neutralization activities	No direct map to strategy
Market Pull	Capability requested by customers	Valid interest by customers	No pull or interest from customers
Complexity and Risk	Low complexity, low risk	Moderate complexity and risk	High complexity, high risk
Effort and Cost	Low effort, cost <$3m	Low effort, cost $3–9m	Large effort, cost >$9m

each of the candidate projects. To facilitate the collection of discussion outcomes and information, it is best to create a work template for use in the voting event. Figure 2.3 illustrates an example Voting Model template that can be used in either physical or electronic format. Some of the most productive prioritization sessions we have witnessed have been those that use a physical voting template that is large enough to hang on a conference room wall.

Voting Model Work Template							
			Prioritization Criteria				
Candidate Projects	Project Priority	Wisdom of Crowds Votes	Criterion 1	Criterion 2	Criterion 3	Criterion 4	Criterion 5

Figure 2.3: Example Voting Model Template

Using the Voting Model

Since Voting Models rely on the expert judgment of a cross-section of stakeholders, it is best if a face-to-face work session be used to ensure proper collaboration and communication takes place. It is within this context that the following steps are recommended for using a voting model.

Step 1: Validate the voting criteria. The initial set of criteria and voting anchor descriptions developed prior to the work session are presented to the stakeholders in this initial step. The intent is to level set the stakeholders on the criteria, how the criteria will be evaluated, and gain buy-in from the cross-section of stakeholders. If necessary, the criteria and voting descriptions can be modified during this step.

Step 2: Review the list of candidate projects. This step involves reviewing all candidate projects that the stakeholders will be evaluating during the prioritization process. To expedite time, the projects can be pre-populated in the voting template. The intent is to ensure all projects are listed and do an initial determination if any of the projects should be eliminated from consideration or if others should be added to the list.

Step 3: Project value proposition and initial voting. Taking each candidate project in the list in order, the project representative describes the project value proposition in three to five minutes. Stakeholders will then have an opportunity to ask clarifying questions and debate the value proposition. It is in this step that an expert facilitator is needed to gage the discussion and debate for focus, value, and time constraints.

At the end of each discussion the stakeholders are asked to vote on the project for each criterion by a show of hands (or electronic vote if blind voting is desired). The project receives a vote of *H, M, L* (or 1, 2, 3) for each criterion as shown in Figure 2.4. This process is repeated for each candidate project on the list.

Voting Model Work Template							
Candidate Projects	Project Priority	Wisdom of Crowds Votes	Prioritization Criteria				
			Criterion 1	Criterion 2	Criterion 3	Criterion 4	Criterion 5
Project 1			M	H	H	L	L
Project 2			L	H	H	M	M
Project 3			L	M	H	L	L
Project 4			L	M	M	H	H
Project 5			L	M	L	H	L
Project 6			H	M	M	M	L
Project 7			L	H	H	M	L

Figure 2.4: Initial Value Voting Assessment Results

Through this step in the process, two significant things are accomplished. First, an initial assessment of candidate project value is debated and scored. Second, the stakeholder's knowledge about the intent and potential value of each candidate project is increased. It is common at this point to want to make a final judgment about project priority. However, doing so creates a fatal flaw that many numeric scoring models fall into. Little separation will still exist between the highest value projects and lowest value projects. To create more separation, one final step is needed.

Step 4: The "Wisdom of Crowds" vote. Tapping into a technique that James Surowiecki termed the wisdom of crowds provides the opportunity to let your new wise set of cross-organizational stakeholders determine the highest value projects.[5] In this step, each stakeholder is given a set number of votes (for example, ten) which he or she can place against the candidate projects. However, each stakeholder has a maximum limit on the number of votes (for example, three) that he or she can place against any one project. This constraint prevents a stakeholder from placing all votes on one particular project. Once again, this voting can take place in the open using the voting model template or electronically if blind voting is preferred.

A final tally of votes is conducted and the top candidate projects are identified. Figure 2.5 shows an example of a fully populated voting model with three bands of prioritized projects based upon the results of the voting. The top band contains the highest value projects which need to be funded and initiated. The bottom band contains

Voting Model Work Template							
Candidate Projects	Project Priority	Wisdom of Crowds Votes	Prioritization Criteria				
			Criterion 1	Criterion 2	Criterion 3	Criterion 4	Criterion 5
Project 1	25	H	M	H	H	L	L
Project 2	19	H	L	H	H	M	M
Project 3	17	H	L	M	H	L	L
Project 4	12	H	L	M	M	H	H
Project 5	11	H	L	M	L	H	L
Project 6	9	M	H	M	M	M	L
Project 7	8	M	L	H	H	M	L
Project 8	8	M	L	H	H	M	M
Project 9	5	L	M	H	H	L	L
Project 10	3	L	L	M	M	H	L

Figure 2.5: Ranking of Projects into High, Medium, and Low Priority

the lowest value projects which need to be eliminated or recast to provide more value. The middle band of projects needs additional scrutiny, but can be funded and initiated if resources are available after all top-level projects are initiated.

Voting Models provide a cross-section of organizational stakeholders in the determination of project ranking which leads to a greater level of buy-in of the final project ranking decision tends to occur (see "A Rock Star Votes").

A Rock Star Votes

Prioritizing a set of technology development projects can be a challenging task. This is especially true in an organization like Intel where a significant number of projects are vying for funding and resources, and each project has an influential and respected technologist championing its value proposition for the company.

Being the person responsible for the final decision about which new technology development projects will be funded and which won't, requires a track record of success, such as the track record of a gentleman named Ajay. Ajay was a co-inventor of the Universal Serial Bus (USB) and featured in one of the companies "Intel Rock Stars" television commercials.

However, as Ajay learned, being an expert in your field and possessing a track record of success does not give you a decision-making mandate. Making project ranking and selection decisions are relatively easy, the hard part comes in gaining organizational buy-in for the decisions you have made. In Ajay's case, two problems consistently emerged once a mandate decision was made:

1. People continued to work on technology projects that were not officially selected and funded, and
2. The product planners who made the decisions on which technologies to include in a project often didn't agree with Ajay's decision, making it difficult for the project to transition from technology development to project development.

Ajay turned to the Voting Model as a potential solution to the problems. He began by inviting both technology experts and product planners to the project prioritization discussions. Through the Voting Model technique, the product planners have an equal say in which technology projects were most valuable (everyone has the same number of votes).

Two other behaviors were also critical. First, Ajay remained quiet during the questioning and debate on project value, and second, he made sure he voted last as to not influence the voting of the other members in the session.

At the end of the session, he had a prioritized project list that was not too far from how he would have personally ranked the projects. But most importantly, he gained alignment with the product planners on which project technologies would eventually make their way into Intel products. Some of them we are now using with our personal computers.

PAIRWISE RANKING

When a small number of candidate projects are involved in the selection process, the Pairwise Ranking technique is an effective tool for prioritizing projects. Pairwise ranking helps to make the complicated process of prioritizing a set of candidate projects simpler by eliminating the need to sort through a list of projects holistically. It forces a structured process and makes the project comparison results immediately visible to the stakeholders and decision makers. Through the project-to-project comparison methodology, only two projects at a time have to be evaluated.

Developing a Pairwise Ranking Tool

Developing a Pairwise Ranking tool is a very simple exercise consisting of two primary steps. First, a comparison matrix must be constructed which represents the number of candidate projects which will be compared. Second, a set of criteria must be identified and documented.

Construct a Comparison Matrix

The construction of the comparison matrix is a simple exercise that is dependent upon the number of candidate projects that will be ranked. Figure 2.6 shows a pairwise comparison matrix for six candidate projects.

Care should be taken to ensure that the numbering schema is accomplished correctly. The numbers representing each of the six projects should be laid out in ascending order from top to bottom in both the vertical and diagonal axes.

Document the Comparison Criteria

A set of criteria, from which each project pair will be compared, is identified and documented to ensure consistency in comparison. These criteria will serve as the basis for comparing each of the candidate projects against one another. Try to limit the criteria to three items, five as an absolute maximum. Anything greater than five criteria adds

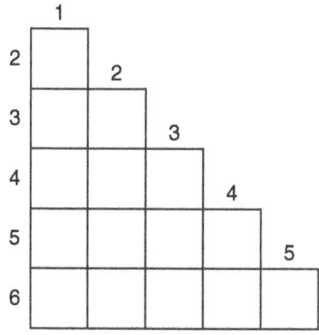

Figure 2.6: Pairwise Ranking Comparison Matrix

unnecessary complexity and makes the tool and comparison exercise more complicated than it needs to be. Pairwise Ranking criteria may include some of the following:

- Net Present Value
- Return on Investment
- Payback Period
- Level of Strategic Alignment
- Cost Savings
- Market Share Increase
- Affordability
- Usability.

Using the Pairwise Ranking Matrix

With a comparison matrix constructed, the comparison criteria identified, the decision method agreed upon, and the right set of stakeholders assembled to compare the candidate projects, the following steps are recommended for using pairwise ranking.

Step 1: Rank each project pair. Systematically work through the candidate projects, comparing one against another, following the structure created by the comparison matrix. For each pair of projects, the stakeholders will determine which of the two candidate projects is preferred based upon the criteria identified. The number of the preferred project is then inserted into the comparison matrix. This project-to-project comparison is repeated until the comparison matrix is completely filled as demonstrated in Figure 2.7.

Step 2: Tally comparison results. Using a simple scorecard (Table 2.6), tally the number of times each candidate project was preferred during the project-to-project comparison exercise. This is accomplished by viewing the inputs in the fully populated comparison matrix.

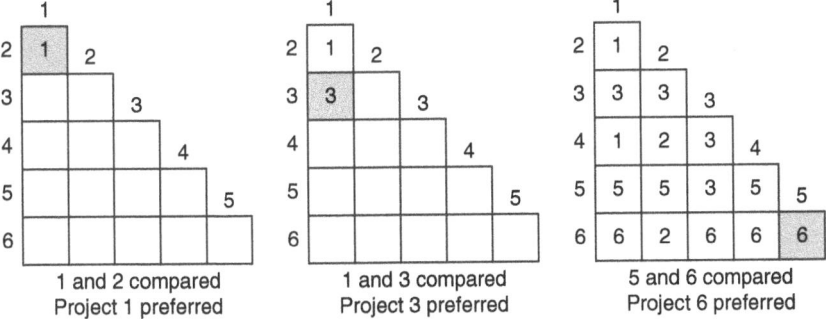

Figure 2.7: Project-to-Project Comparison

Table 2.6: Project Preference Tally						
Candidate Projects	1	2	3	4	5	6
Times preferred	2	2	4	0	3	4

Table 2.7: Candidate Project Ranking						
Candidate Projects	1	2	3	4	5	6
Times preferred	2	2	4	0	3	4
Project ranking	**4**	**5**	**2**	**6**	**3**	**1**

Step 3: Rank the candidate projects. Using the tally information from the previous step, rank the candidate projects based upon the number of times they are preferred. Table 2.7 illustrates the priority ranking of the six projects used in this example.

In cases where there is a tie between two candidate projects, refer to the comparison matrix created in step one and find the box where the two projects were compared against one another. The project which was preferred between the two receives the higher ranking. In the example above, projects 3 and 6 were each preferred four times. By referring back to Figure 2.7, one can see that when the two projects were compared against one another, project 6 was preferred. Therefore, project six receives the higher ranking.

Benefits

Pairwise ranking helps to make a complicated process (prioritizing a set of candidate projects) much more simple by eliminating the tendency to try to sort through a list of projects holistically. Through the project-to-project comparison methodology, only two projects at a time have to be evaluated.

The tool itself forces a structured process and makes the project comparison results immediately visible to the stakeholders and decision makers. The tablature nature of the final ranking results is also more visibly effective than other project scoring and ranking tools.

ALIGNMENT MATRIX

The project Alignment Matrix is prepared for project portfolio management reviews when projects are assessed with respect to alignment with the strategies, goals, and objectives of an organization. It establishes the degree to which a project is aligned with the organization's business strategy (see Figure 2.8).

Based on the alignment information, the preliminary selection and risk balancing of projects may be changed, and the final project portfolio adopted, providing improved alignment of projects with the strategic goals of the organization.

Developing the Alignment Matrix

The first column of the matrix contains the list of the organization's business strategic goals that serve as criteria to align projects with the organization's strategy. Then, in each of the remaining columns, the degree of alignment of individual projects with each goal

Example Business Strategies	Project 1	Project 2	Project ...	Project n
Provide clearly differentiated products from our competitors	P	F		P
Consistently deliver performance increase in high-speed devices	F	N		P
Be the first to market with new products	F	F		F
Supports the common platform architecture	F	F		F
Continuously reduce manufacturing cost	N	F		P
Legend: F = Fully Supports P = Partially Supports N = Does Not Support				

Figure 2.8: Example Alignment Matrix

is assessed using a qualitative scale. As an outcome, a qualitative goal-by-goal alignment evaluation for each project is generated that may be used for different strategic purposes.

The assessment of a project's alignment with an organization's business strategy calls for information that typically comes from three inputs:

1. Approved business strategy
2. The portfolio of projects
3. Project business case (or preliminary business case information).

The approved business strategy provides a list of the organization's business goals that the strategy is striving to accomplish. To assess the degree of alignment of individual projects to the strategic goals, a list of current and future projects is needed. This information is typically found in project portfolio documents. Finally, to understand how well each project is aligned with the strategic goals, the preliminary project business case is needed.

Identify the Organizations Strategic Goals

Strategic goals are defined by the organization's senior management, sometimes formally in strategic plans, at other times informally. In either case, the goals should be used for the Alignment Matrix assessment. Since each organization is a unique entity, the list of strategic goals found in the Alignment Matrix will be unique as well. Additionally, as the strategic goals of an organization are updated and modified, the list of goals in the Alignment Matrix needs to be updated accordingly.

Identify the Projects

There are two steps to this action. First, the names of the new and existing projects are entered into the columns of the Alignment Matrix. As stated earlier, the list of the projects is normally part of the portfolio of projects documentation. If a formal portfolio of projects does not exist, an active project roster will suffice. Second, the project strategy and goals should be developed and documented to secure the information needed to assess alignment of projects to strategy.

Define the Alignment Scale

Scales vary and the choice of scale to use is organization specific. We believe that a simple, qualitative scale is completely adequate and provides the value we want from this matrix. An example of a simple qualitative scale is three-level scale shown in Figure 2.8. The scale includes Fully Supports—for the highest degree of alignment, Partially Supports—for the medium-level alignment, and Does Not Support—for no alignment of the project with a specific goal.

Assess the Degree of Alignment

Now it is time to assess each project's alignment with each organizational business goal using the adopted scale of assessment. Typically, decision makers who do the assessment should use a collaborative work session format, where information from multiple perspectives is exchanged, and the assessment decisions are shared and consensual.

Variations

An infrastructure technology (IT) organization for a major financial institution uses a variation of the Alignment Matrix, which they call the strategy alignment map (Figure 2.9) to gain alignment between the IT strategy and business strategy. The map

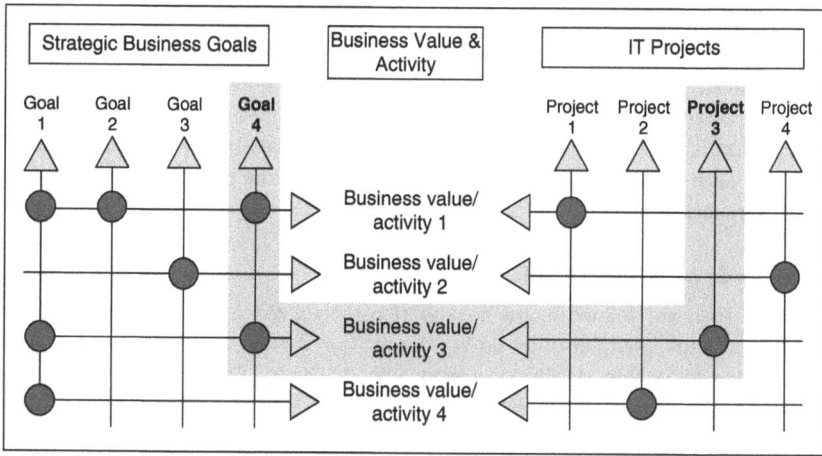

Figure 2.9: Strategy Alignment Map

is used to ensure their IT projects complement the needs of the business. Their IT PMO director Melida Ramos explains, "The alignment chart provides a strategic mapping of the business goals, the business values, and the IT projects. As part of the strategic changes which occurred last year, we were tasked by the company to design the alignment process, part of which was accomplished by the alignment map which helped us visualize the alignment between business goals, business value and the projects."

A bubble on the intersection of strategic business goal 4 and business value 3 means that they are aligned. Further, a bubble indicates that business value 3 intersects with project 3, meaning they are aligned. In summary, project 3 delivers business value 3, which helps to achieve strategic business goal 4. The alignment is about IT projects contributing to business benefits by helping to achieve the strategic business goals.

Benefits

Typically, the Alignment Matrix is prepared for project portfolio management reviews when projects are assessed with respect to alignment with the strategies, goals, and objectives of an organization. Based on the alignment information obtained by means of the Alignment Matrix, the preliminary selection and risk balancing of projects may be changed, and the final project portfolio adopted, providing improved alignment of projects with the strategic goals of the organization.

The Alignment Matrix enables an organization to refine the selection and risk balancing of the preliminary portfolio of projects by pointing to projects that are best aligned with the organization's business goals. Based on that, one can eliminate some preliminary selected projects that are not strategically aligned; one can also add new projects that are better aligned. That is the matrix's value, which is strategically precious given how difficult it is to select the most valuable projects that are risk-balanced and are also aligned with the strategic goals of the organization.

Additionally, it requires that senior managers of an organization to develop and document the organization's strategic goals. The Alignment Matrix aids the project manager in understanding how well his or her project supports the strategic objectives of the firm, and therefore helping him or her in creating the project strategy.

PROJECT BUSINESS CASE

The Project Business Case, sometimes called the project proposal, is used to assess the feasibility of a project from multiple business perspectives. It demonstrates how the project will contribute to business results and how the project aligns with the strategy of the organization.

The Project Business Case describes a business opportunity in terms of alignment to strategy, market or customer needs, technology capability, and economic feasibility. It also provides a balanced view of business opportunity versus business risk. The Project Business Case is used to help top managers make sound decisions when considering investment options. It establishes alignment between strategic business goals and the project. When used consistently, it provides a consistent set of criteria to evaluate projects against others in the portfolio.

The intended outcomes of the Project Business Case is :

- To gain agreement on project scope and business success criteria
- To obtain approval of funding and resource allocation for project planning and implementation
- To obtain approval to proceed from origination to planning the project

Developing The Project Business Case

The business case for a project must be brief, unbiased, and clear. This requires quality information about the following:

- *The business environment.* Information about the business environment is used to position and differentiate the product, service, or infrastructure capability being proposed.
- *The business strategy.* The business strategy specifies the strategic goals the organization is striving to achieve through the project.
- *Business success criteria.* Success criteria indicate the goals and metrics that the project must achieve to be considered a success.

Table 2.8 can be used as a guide for developing a Project Business Case. It suggests a minimum set of information to include in the business case.

Table 2.8: Minimum Elements of a Project Business Case	
Business Case Element	**Description**
Executive summary	A summary of the business case that gives stakeholders a brief overview of the project
Background information	A brief description of the environment and business context of the project
Project purpose	A succinct statement of the problem to be solved or the opportunity to be realized. The project purpose should state how the project aligns with the organization's strategic plan
Value proposition	A succinct statement characterizing the value and/or benefits to be delivered (quantified when possible)
Market assessment	For commercial projects, an overview of the market, competitors, legal, environmental, and competitor information
Project objectives	The objectives the project is expected to achieve along with quantifiable measures of success
Financial analysis	Key financial indicators such as payback time, net present value, and internal rate of return
Critical assumptions	The events and circumstances that are expected to occur for successful realization of the project objectives
Risk analysis	A high-level assessment of the risks which may prevent realization of the objectives and business benefits

Describe the Business Opportunity

The business opportunity is described in the background, purpose, value proposition, and market assessment. The information in these sections of the business case provide a well-rounded understanding of what the project will deliver and what the organization can expect to gain by investing in the project.

Define Business Success

Business success is defined by listing the project objectives (i.e., the goals to be achieved), and the quantifiable measures that indicate success in achieving the objectives.

For commercial endeavors (as opposed to projects for the public good), financial objectives are of primary interest to senior management. Financial criteria can include payback period, net present value, and internal rate of return as described previously. This information is often used to develop a cost/benefit analysis. The cost/benefit analysis should identify both tangible and intangible benefits, with the benefits expressed in quantifiable terms such as dollars gained or saved, hours saved, and increase in gross margin.

Identify Critical Assumptions

Much of the work performed when evaluating a project is focused on trying to predict what will happen in the future. To do this, a series of assumptions about future events and circumstances are made. By explicitly stating the primary assumptions the project business case is built on, the project manager establishes a common understanding of how the future may unfold.

Assess Project Risk

At this point in the project only high-level risk events and conditions are known. These should be described along with the means of minimizing the threats to achieving project objectives and success criteria. Where possible risk assessment should estimate the probability of project success and an assessment of whether the level of risk warrants continuing the investment in the project.

Using the Project Business Case

Presentation of the business case to top management stakeholders is normally used to drive the final investment and funding decision near the end of the origination stage. Information in the business case provides primary content for the Project Charter (the Project Charter is described in an upcoming section).

The Business Case is first developed prior to project startup; however, it should be revisited throughout the project and considered a living document which can be updated as more accurate information becomes known.

During the project the business case may be reviewed periodically, such as prior to releasing funds for large expenditures. The business case should be revisited during the project retrospective to evaluate if the project was successful in achieving the intended business goals.

References

1. De Wit, B. and Meyer, R. (2010). *Strategy: Process, Content, Context*, 4th ed. London: Cengage Learnings.
2. Pearce, J.A. II and Robinson, R.B. Jr. (2008). *Strategic Management: Formulation, Implementation, and Control*. New York: McGraw-Hill.
3. Martinelli, Russ and Jim Waddell. "Aligning Program Management to Business Strategy." Project Management World Today, January–February 2009.
4. American Productivity and Quality Center. New Product Development: Embracing an Adaptable Process, Final Report, APQC, 2010.
5. Surowiecki, J. (2005). *The Wisdom of Crowds*. Harpswell, ME: Anchor Publishing.

3

PROJECT ORIGINATION

One of the main reasons projects fail is unclear and/or poorly communicated project objectives. After all, if people don't understand the purpose of the project, the intended outcomes, and what success looks like, there is very little chance they will deliver what is expected. Conversely, a project that is well understood, with clear outcomes and objectives, has a much higher likelihood of achieving success.

Project origination is concerned with ensuring that there is a consistent high-level understanding of what a project is intended to achieve and how the outcome will be accomplished. This includes ensuring that the financial, capital, and human resources are committed to the project, that the objectives are adequately defined, and the level of complexity is understood.

Because one size does not fit all when it comes to project origination we will look at several different documents you can use at the start of the project to ensure a common understanding. A complexity assessment can be used at the portfolio level to ensure an appropriate balance of projects in a portfolio. It can also be used at a project level to manage expectations. Complex projects contain more uncertainty and risk, and therefore require higher-skilled resources and a more robust contingency fund.

A Project Charter and a Project Canvas are two means of capturing high-level information about a project. A Charter is often used on large projects, but may be too much for a smaller or simple project. A Project Canvas works for smaller projects, or to provide a summary-level overview of the project that can be elaborated in the Charter.

A Project Roadmap identifies high-level information about how and when the work will be accomplished. It shows project phases, milestones, key deliverables, and phase gates.

An Assumption and Constraint Log is started at the beginning of the project. It is a dynamic document that is updated throughout the project lifecycle. These documents can be used in whatever combination suits the project and organizational needs.

Project Management Toolbox: Tools and Techniques for the Practicing Project Manager,
Third Edition. Cynthia Snyder Dionisio, and Russ J. Martinelli.
© 2025 John Wiley & Sons, Inc. Published 2025 by John Wiley & Sons, Inc.

COMPLEXITY ASSESSMENT

Complexity is a function of multiple interactions that can occur with human behavior, systems, and cultures. Complexity often has an element of ambiguity to it that makes it difficult to understand the influences and the outcomes of interactions. It is a characteristic of many projects. There are several contributing factors to complexity. Consider the following:

- Designs have become more complex as features and integrated capabilities increase
- The process to develop and manufacture solutions is requiring more partners, suppliers, and others throughout the value chain
- The ability to integrate multiple technologies with end user wants requires not only accuracy regarding requirements delivery, but also speed and agility
- The current global, highly distributed business environment requires work to occur in multiple sites across multiple time zones
- The use of artificial intelligence is increasing, which can simplify routine tasks, but adds a layer of machine/human interaction that has not been present before.

Therefore, the ability to characterize and profile the degree of complexity associated with a project has become more common for executive leaders, portfolio managers, and project managers.

The information gained from the Complexity Assessment helps to balance the portfolio of projects with an appropriate mix of high-, medium-, and low-complexity projects. For a project manager, it aids in the determination of the skill set and experience level required of the project team; guides the implementation of key project processes such as change management, risk management, and contingency reserve determination; and helps the project manager adapt his or her management style relative to the level of complexity.

Developing the Complexity Assessment

The structure of the project Complexity Assessment includes four steps.

1. Identify the various dimensions you want to assess.
2. Establish a scale for each dimension with the low and high end defined.
3. Score each dimension.
4. Connect the scores of each dimension with a line to establish a complexity profile.

The complexity profile is a graphical representation of a project's multifaceted complexity. An example of a project complexity assessment is illustrated in Figure 3.1.

Determining the Complexity Dimensions

Every industry has unique characteristics, every business within an industry is unique, and every project within a business is unique. This means that a firm has to customize the complexity assessment tool for its use.[1] Project managers often start this work by determining the complexity dimensions that are specific to the organization. Table 3.1 shows an example of various factors that make up technical, structural, and business complexity.

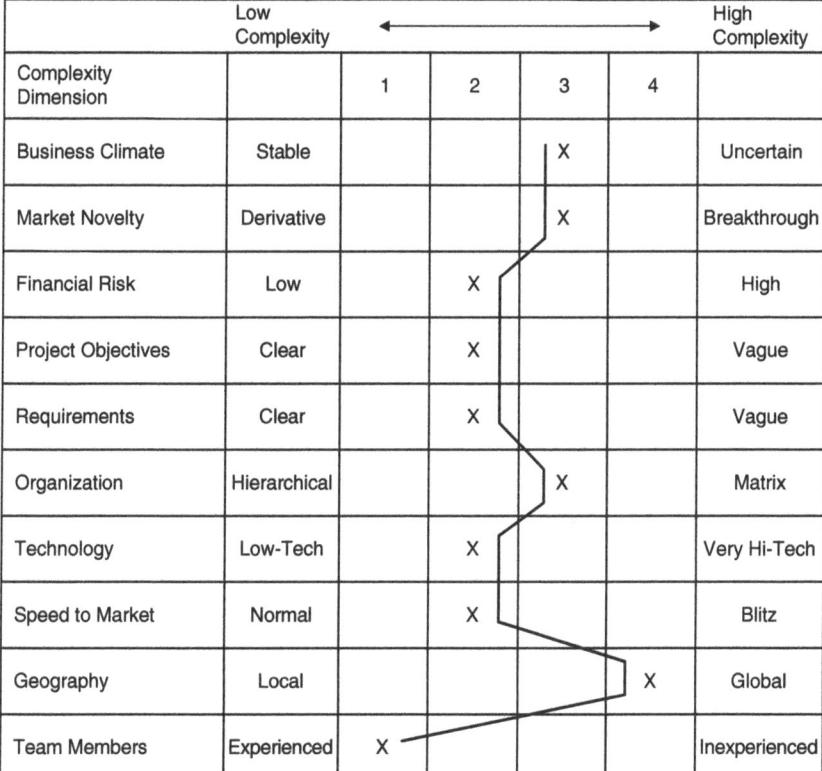

Complexity Dimension	Low Complexity	1	2	3	4	High Complexity
Business Climate	Stable			X		Uncertain
Market Novelty	Derivative			X		Breakthrough
Financial Risk	Low	X				High
Project Objectives	Clear	X				Vague
Requirements	Clear	X				Vague
Organization	Hierarchical			X		Matrix
Technology	Low-Tech	X				Very Hi-Tech
Speed to Market	Normal	X				Blitz
Geography	Local				X	Global
Team Members	Experienced	X				Inexperienced

Figure 3.1: Example Project Complexity Assessment

Define the Complexity Scale

With the dimensions of complexity identified the next step is to define how each dimension of complexity will be measured. You do this by choosing a scale for each dimension. In the example above the complexity for each dimension is measured on a simple scale with 1 being the lowest complexity and 4 being the highest complexity. The important thing to establish is not so much the scale, but rather the anchor statements for the scale. For example, a low complexity rating for requirements would have clear requirements, whereas a high complexity rating would have vague requirements.

Anchor statements help build consistency when assessing the complexity for each dimension. Without anchor statements each assessor may evaluate the levels differently, leading to inconsistent complexity evaluations. Well-defined anchor statements help to ensure all assessors approach the scale for each complexity dimension from a consistent frame of reference.

Using the Complexity Assessment

Once the complexity dimensions are identified, each dimension is then assessed based upon the established scale. For example, in Figure 3.1, speed to market is assessed as a level 2 complexity.

Table 3.1: Example Technical, Structural, and Business Complexity Dimensions	
Technical Complexity	
Low Complexity	**High Complexity**
Feature upgrade to an existing product	New product architecture and platform design
Development of a single module of a system	Development of a full system
Use of existing and developed technologies	Use of new and undeveloped technologies
Structural Complexity	
Low Complexity	**High Complexity**
Team is co-located	Team is a geographically distributed
Mature processes and practices	Ad hoc processes and practices
High performing team	Low level of team cohesion
Single-site development	Multi-site development
Single-geography development	Multi-geography development
Single-cultural team	Multi-cultural team
Single-company development	Multi-company development
Business Complexity	
Low Complexity	**High Complexity**
Selling into traditional and mature markets	Selling into new and emerging markets
Receptive customers and/or stakeholders	Unreceptive customers and/or stakeholders
Flexible time-to-money requirements	Aggressive time-to-money requirements
Existing end-user usage models	New end-user usage models

Once all complexity dimensions are assessed, connect the obtained scores for each dimension to produce the complexity profile to help visually depict the overall project complexity. The profile in Figure 3.1, for instance, indicates that the program is of medium complexity, with all dimensions at levels 2 and 3, except team members who are experienced (the least complex) and a globally distributed team (the most complex).

Typically, the project Complexity Assessment tool is prepared very early in the project cycle. However, this tool should be utilized dynamically and updated periodically in high-velocity environments where the project scope and business climate may change frequently.

By using this tool, the senior management team and project manager can quickly get a feel for the level of complexity of each project within the organization. However, care should be taken to prevent the inclusion of too many complexity dimensions. In this case, the simpler the structure of the tool, the more effective its use will be.

PROJECT CHARTER

The Project Charter is a document that formally authorizes a project. It defines the reason for the project and assigns a project manager and their authority level. This is especially important in environments where project managers have no direct authority over project team members and other resources but bear the responsibility for delivery of the project outcome. In such a situation, for the charter to be effective, the issuing manager must be on a level commensurate with the type, number, and cost of the resources needed for the project that has control over the resources.

Typical contents of a charter include:[2]

- Project purpose
- High-level description
- Key deliverables
- High-level requirements
- Project objectives and related success criteria
- Overall project risk
- Summary milestone schedule
- Summary-level budget
- Assigned project manager and authority level
- Sponsor information
- Project approvals

Developing a Project Charter

The Project Charter serves as an agreement or contract between the project manager and the executive sponsor to work as a team to execute the project to the intent of the Project Business Case, and to utilize the organization's resources to maximum business benefit.

When comparing the type of information described in the charter and the business case, you will note some similarities. Both contain the project purpose, objectives, and success criteria, for example. Where these elements differ is their level of detail. More precisely, because it is an authorization tool, the Project Charter tends to include more detail than the business case because it provides the information needed by the project team to proceed with the project.

Define the Project Purpose

The project purpose describes why the project is being authorized. The purpose may be carried over from the business case. It generally ties the project to the organization's strategic goals. For example:

- Reduced costs
- Increased customer satisfaction
- Improved efficiency
- Entering a new market
- Increased market share.

The purpose is bigger than the project outputs, such as improving an internal process or launching a new product, it identifies the intended outcomes, rather than the outputs of the project.

High-level Project Description

The project description identifies both the project and the product scope description. The product description identifies what will be delivered. The project description identifies the project work it will take to deliver the product(s). An example of a product description could be information about a new manufacturing facility. An example of the project description could include information on research, procurement, testing, documentation, training, and so forth. The project work supports and enables the product work.

Key Deliverables

Key deliverables should include both project and product deliverables. The product deliverables are entirely dependent on the individual projects. Project deliverables typically include the project schedule, budget, project management plan, along with any specific subsidiary management plans, such as a risk management plan, training management plan, or other relevant supporting documentation.

When a project is done under contract, having clear and precise descriptions of both product and project deliverables is critical. While the charter does not have the same legal importance as a contract, it does provide a foundational agreement about the project.

High-Level Requirements

The charter documents requirements at a high level. It includes the high-level conditions or capabilities that must be met to satisfy the purpose of the project. The requirements describe the product features and functions that must be present to meet stakeholders' needs and expectations. These will be further elaborated in requirements documentation.

Project Objectives and Related Success Criteria

Project objectives are the goals you want to achieve for the project. They are usually established for at least scope, schedule, and cost; though some organizations include quality, safety, and stakeholder satisfaction objectives as well.

The success criteria identify the metrics or measurements that will be used to measure success in attaining the objectives. Success criteria for schedule would be meeting the project delivery date, for budget, it would be coming in at or under the authorized budget, and so forth.

Stretch Goals, Or Not?

How attainable should the project goals in the charter be? Is it okay to write charters using a stretch goal? Empirical evidence suggests that those who set stretch goals—that is, outline goals that are typically quite difficult to attain—outperform those with routine goals, which are typically easy to attain. If you are purely driven by performance, the choice is clear: go for stretch goals.

At Google, many project managers deliberately use stretch goals in their project charters because corporate culture drives this behavior. What happens when they do not attain their stretch goals? According to one project manager, "No project sponsors use this practice to call people out. The idea is to always strive for more and do your best. If you do so, you won't be penalized if a stretch goal is missed."

Is this the case in all companies? According to a project manager for a business-to-business software company, "If you try stretch goals and miss, it is likely to be held against you in your next performance evaluation. That's why everybody goes for routine goals in our company." The point, then, is that the use of project stretch goals is dependent upon company culture and likely the level of desire for industry leadership.

Overall Project Risk

Overall project risk is different than individual risk events. It is an assessment of the overarching uncertainty surrounding the project. This may include political or price volatility, ambiguity about the market or competition, or complexity associated with the project or project solution. Overall project risk may not be relevant for smaller, shorter-term projects. However, as the size, cost, duration, and complexity of the project increase, so does the need for addressing overall project risk.

Summary Milestone Schedule

A milestone is a significant event in the project, such as the completion of a key deliverable, the beginning or end of a project phase, or product acceptance. The charter should only identify the significant milestones. Specifying five to eight milestones ensures you are staying at a summary level, rather than getting into too much detail.

Summary Budget

The summary budget identifies the funding for the project. The summary budget may include sources of funding and annual funding limits. It may also include information on fixed and variables categories of costs, contingency reserve, and management reserve.

Assigned Project Manager and Authority Level

Depending on the organization's culture and structure, the project manager may have total authority, or very little authority. It is a good practice to identify the authority and limits to authority for at least budget, staffing, technical decisions, and conflict resolution.

Budgetary authority refers the ability to commit funds and manage the budget. In addition to identifying the project manager's budget authority, the variance threshold should be documented. This communicates the variance level above which the project manager must escalate cost issues.

Staffing authority relates to the ability to identify, accept, or reject project team members. In some organizations the project manager accepts whichever team members are assigned. In other organizations the project manager has the authority to choose their team members, including outsourcing, hiring, disciplining, and firing staff.

Technical decisions refer to the authority to make decisions about product components, deliverables, and processes for the project.

Most projects face some degree of conflict at one time or another. The authority to resolve conflict within the team, within the organization, and external to the organization should be documented to ensure everyone is clear on the approved levels of conflict resolution.

Sponsor Information

One of the purposes of issuing a charter is to formally announce the project, the project manager, and the sponsor. The specific role of the sponsor varies by organization. However, sponsors typically provide guidance to the project manager, act as a champion for the project at the management level, and provide political cover for the project team.

The sponsor is a senior manager who has authority over budget and resources, especially when the project manager does not have that authority. If the project manager has authority to make budget or resource decisions up to a certain amount, the sponsor is the person who has the authority above that limit.

Additional types of authority include the ability to approve significant changes, determine acceptable variance limits, resolve inter-project conflicts, and work with functional managers in the case of team member issues.

Project Approvals

As an authorization document to officially start expending organizational resources, the Project Charter must include the names, titles, and signatures of the individuals who will sign-off on the project. At a minimum, signatures should be required for the project sponsor and the project manager.

Variations

The practice of project chartering exhibits many variations and nuances, including its name, content, pattern of use, and formality. For instance, some organizations call it the "project initiating document," others the "project brief." In all cases, the charter is meant to bring a project into existence.

As for the content, some organizations use charters that include specifics about budget and schedule for major milestones, as shown in Project Charter example in Figure 3.2. Others, especially for smaller projects, find it sufficient to announce the purpose of the project, the start of the project, the objectives and acceptance criteria, along with the project manager.

Project Name:	ISU Alumni Website
Project Manager:	Jen Cosgrove
Project Sponsor:	Dan Seales

Project Mission

This project will provide the local chapter of the ISU Alumni Association a web site that will be used as a resource to enable continued social networking, information exchange, and information repository for alumni association members.

Project Goals

The university is looking for new ways to help alumni stay connected to the university postgraduation. The web site to be developed will (1) create the means to establish a strong alumni social network; (2) provide a portal for the university to communicate activities, information, and needs; and (3) establish a repository of academic research information for the alumni to access and contribute to. The project will be completed prior to the Fall 2017 academic semester, and cost no more than $60,000 to implement.

Project Scope

The scope of the project will include the design, development, test, and go-live activities necessary to create an operational web site. The web site will include four major capabilities: (1) HOME page for alumni information and navigation, (2) SOCIAL NETWORKING page, (3) ACADEMIC RESEARCH repository, and (4) ALUMNI ACTIVITIES page.

Dependencies, Risks, and Assumptions

Dependencies	Risks	Assumptions

■ Budget approval
■ Availability of
  IT resources
	■ First use of open
	  source software
	■ Schedule is aggressive
		■ Additional security software is not
		  needed
		■ Open source software can
		  be leveraged

Major Milestones:

Usage study completion	Operational test completion
Web site design completion	Go-live launch completion
Prototype development completion	30-day retrospective completion
Development completion	

Sponsor and Project Team

Project Sponsor:	Dan Seales
Project Manager:	Jen Cosgrove
User Exp. Designer:	Lynda Carmody
Web Site Developer:	Ajit Chattergee
Web Site Developer:	Fariba Rezzanie
Quality Assurance:	Will Torday

Budget and Completion Date

Budget:	$60,000
Completion Date:	August 1, 2017

Approvals

Project Sponsor: _____

Project Manager: _____

Finance Manager: _____

Figure 3.2: Example Project Charter

> ## The Need for the Charter
>
> Do you need a charter for all projects? Consider that it normally takes a leading truck manufacturer located in the U.S. months of work to issue a charter for a new truck development project. With millions of dollars involved, the company develops multiple scenarios of scope, cost, and timeline, and evaluates them carefully before launching the effort. The launch begins with the issuance of a detailed charter, where typically the sponsor is a corporate vice president.
>
> In contrast, a major information technology upgrade project within the same company typically starts with a few sentences, e-mailed to the functional mangers providing resources.
>
> This is a good example to review when deciding whether you need a charter for all projects. The need for the charter should be matched with the size, complexity, and the degree of cross-functional involvement on the project. For projects that don't need a charter, but do need more than a few sentences, a Project Canvas is a good option.

PROJECT CANVAS

Many project managers think that the only way, or the best way to initiate a project is by developing a project charter. A charter is a great way to document information for a large, long-term project. However, the majority of projects don't require a multi-page narrative description. Projects that are less than 6 months can generally get by with a lighter weight means of providing a high-level summary of the project.

A project canvas is a high-level visual summary of a project. It is a framework that allows you to capture new project information quickly and succinctly. In contrast to a project charter, the project canvas is designed to fit on one page. While a one-page summary may not be appropriate for large or complex projects, it is a useful way to document all the key information about a project in an easy-to-absorb format.

The project canvas was developed by Antonio Nieto-Rodriguez and is modeled after the business startup canvas and the lean startup canvas. Unlike those documents, the project canvas focuses on key project information rather than market, competitor, and distribution information.[3]

Developing the Project Canvas

Because the project canvas is only one page, you will want to limit the contents to just the necessary information. Additionally, rather than writing in full sentences, most of the information will be in bullet points.

The sample template in Figure 3.3 shows a tailored version of the original project canvas.

The sample template includes seven fields. These are generic fields that can be customized based on the type of project.

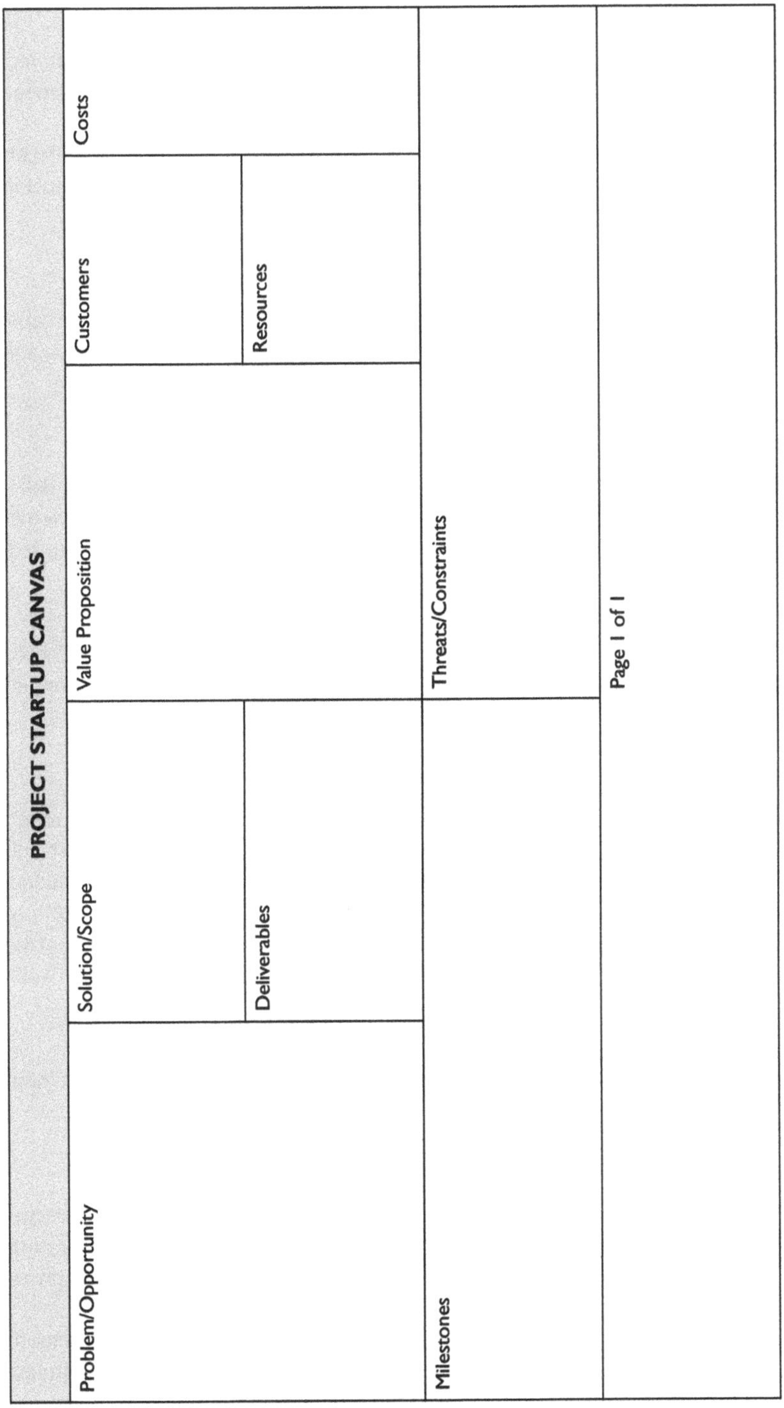

PROJECT STARTUP CANVAS				
Problem/Opportunity	Solution/Scope	Value Proposition	Customers	Costs
	Deliverables		Resources	
Milestones		Threats/Constraints		
Page 1 of 1				

Figure 3.3: Template for a Project Canvas

Problem or Opportunity

Projects are originated to address problems or opportunities. The first field for a project canvas is to identify the problems the project will solve or the opportunities it will meet. Some organizations may choose to call this the project purpose or mission.

It is helpful to include some brief contextual information, such as why the project is needed at this time or if there are external or regulatory factors that are influencing the project.

Solution or Scope

This section describes the solution to the problem or opportunity, or the project scope. It is useful to articulate what is considered in scope and out of scope to help manage stakeholder expectations.

Value Proposition

The value proposition is a concise statement of the value that the project and its deliverables will provide to stakeholders. Stakeholders may comprise internal or external customers. This section is where you will describe why the project or product is needed.

Key Deliverables

Key deliverables can include project and product deliverables. The deliverables should be kept at a high level in the project canvas, though they may be further elaborated in the WBS.

Costs

The cost section provides an initial cost estimate for the project. This is usually a rough order of magnitude estimate that will be further refined when more detailed cost estimates and the project budget are developed. Generally only project costs are included, not operating costs. If required you can differentiate between fixed and variable costs, or in-house and outsourcing costs. If you have an idea about the amount or percentage of the budget that will be used for contingency funds you can include that information as well.

Milestones

List up to five milestones. The milestones in the canvas can be used in the project roadmap and for building the schedule.

Required Resources

Information on resources may identify the staff team members as well as any contractor or vendor resources. If the project requires significant physical resources you may want to split the "resource" field in two so that team resources have their own field and physical resources are listed separately.

The project canvas is easily customizable. You can insert, delete, or combine fields to capture the information you need. For example, a new product development project could include distribution channels, customer segments, and revenue streams rather than deliverables, resources, and milestones as shown in Figure 3.3.

PROJECT ROADMAP

The project roadmap is a high-level graphical summary of the project. The purpose is to present a one-page visual depiction of the project life cycle along with key events and deliverables. It shows the start and end dates along with phases and the flow of the work.

The project roadmap often serves as a transition from the charter or project canvas to the schedule. The charter and canvas list milestones, the roadmap shows how they relate to the deliverables and other work being done, and the schedule develops the detail associated with performing the work.

At a minimum a project roadmap should include the following information:

- Project life cycle phases
- Major deliverables
- Milestones
- Timing and types of reviews.

Developing the Project Roadmap

Developing a project roadmap requires upfront thinking about the project, the deliverables, how you plan on creating the deliverables, the governance, and high-level dependencies. There are five steps to create a project roadmap:

1. Determine the governance needs of the project
2. Identify the project life cycle
3. Identify the major deliverables
4. Determine key milestones
5. Establish the timing of significant reviews.

The first step is to determine the governance needs of the project. This entails working with the sponsor to assess how much oversight is needed for quality assurance, contractual compliance, regulatory compliance, and customer acceptance. An internal project with no regulatory or safety issues will have less governance needs than a project done under contract that entails system integration, regulatory requirements, and multiple sign offs. This information will influence phase gates, reviews, and dependencies between project phases.

The second step is to determine the phases in the project life cycle. If your organization performs many similar projects, you may have an established life cycle with phases and control gates. If not, you will consider the type of work being done and identify the phases associated with the work. A project for building a new system has different types of work, and therefore different phases than a project to put on a music festival.

As part of identifying the life cycle phases you should consider the nature of the work along with the governance needs. This information will determine if you can overlap phases, or if you need to have a formal phase gate review and sign off before moving forward.

Next the major deliverables are indicated and associated with the applicable phases along the timeline. Usually, deliverables are indicated with a shape or an icon, such as a triangle or diamond, to minimize the amount of text on the roadmap.

Projects that have one main deliverable at the end will be different than projects that have multiple deliverables throughout the project. If one or more of your deliverables are using adaptive methods, you will need to estimate when the deliverables will be complete. If you have multiple deliverables that use different development approaches (such as waterfall and Agile), it may be useful to create swim lanes to show the governance associated with waterfall work and the releases associated with the Agile work.

Next enter in the milestones. Milestones are usually indicated with a different shape or icon from the key deliverables. Milestones may be associated with the start of finish of phases, the kickoff meeting, customer acceptance, and other significant events.

The final piece of information is key reviews. Reviews reflect the governance needs and the type of work. A system development project may have baselines reviews, design reviews, and integration reviews. A construction project may have an architecture review, a walk through, and an occupancy review.

Figure 3.4 shows a sample project roadmap for a nine-month project with phases, phase gates, deliverables, milestones, and reviews.[4]

ASSUMPTION AND CONSTRAINT LOG

Assumptions are factors that are considered to be true but without proof. An Assumption and Constraint Log is useful in ensuring that assumptions are visible and helps to get all stakeholders in alignment so as to minimize conflict and unfulfilled expectations.

In the beginning, most of the work is based on assumptions, such as resource availability, cost estimates, and durations. An assumption log is a centralized place to keep the assumptions up to date.

Constraints are limiting factors that affect the execution of the project. Typical constraints include a predetermined budget or fixed milestones for deliverables. Documenting this information in a log makes them visible and easier to track and validate as the project progresses.

Developing the Assumption and Constraint Log

The Assumption and Constraint Log can be developed in a simple spreadsheet. This allows the project manager to sort on category, status, or due date. Table 3.2 shows the information documented in an Assumption and Constraint Log.

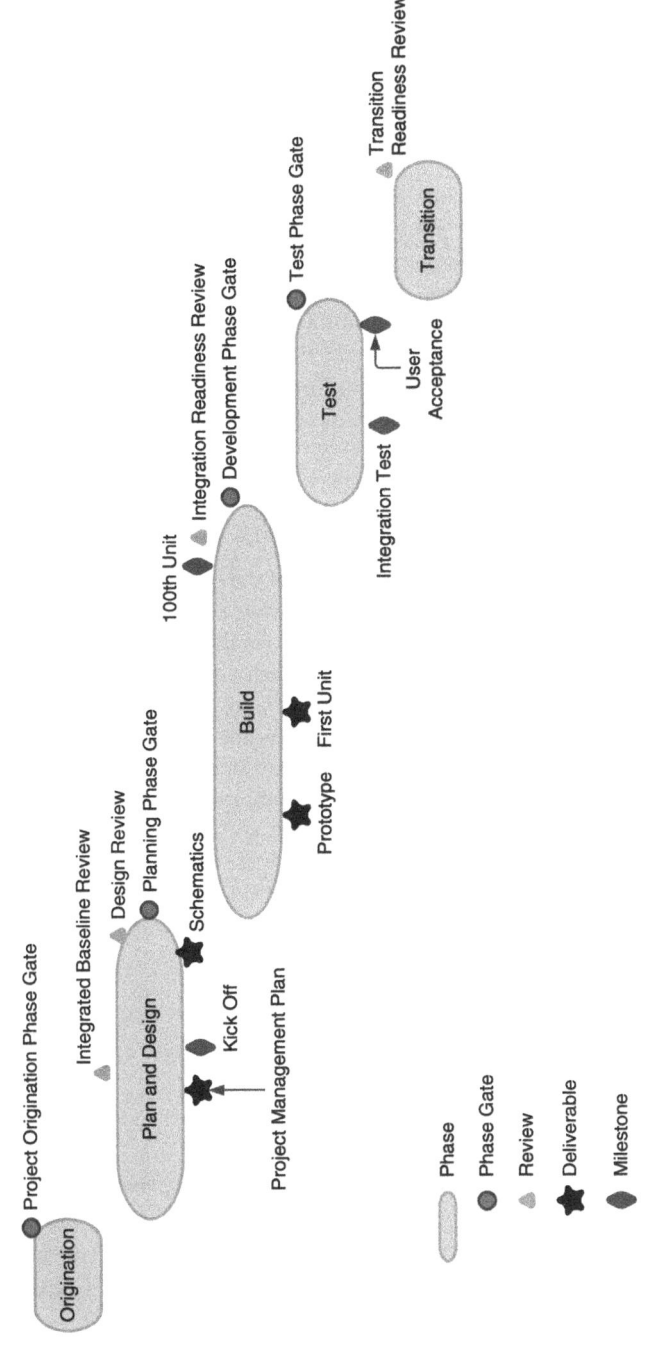

Figure 3.4: Sample Project Roadmap

Table 3.2: Contents in an Assumption and Constraint Log

Document Element	Description
ID	A unique identifier, often alpha numeric with an "A" for assumptions and a "C" for constraints.
Category	The category of the assumption or constraint, such as cost, resource, and technical.
Assumption/Constraint	A narrative description of the assumption or constraint.
Responsible party	The person who is tasked with following up on the assumption to validate if it is true or not.
Due date	The date by which the assumption needs to be validated.
Actions	Actions that need to be taken to validate assumptions, such as confirming start and end dates for resources, conducting market research, or other relevant actions.
Status	The status of the assumptions, such as active, transferred, or closed.
Comments	Any additional material that will provide clarity or relevant information the assumption or constraint

Using the Assumption and Constraint Log

Assumptions can be put forth by any member of the project team. They often surface while creating the origination documents and while planning the project. Though, in practice, they can occur at any time. Constraints are generally known up front and are often indicated in the charter or canvas, such as the budget or delivery date. Constraints may be determined by the customer, sponsor, or regulatory agencies.

During project origination, all activities and planned outcomes are predicated on a set of assumptions of how the future will play out. Since the future rarely goes as predicted, some of the assumptions made at this early stage will be incorrect. This means that the assumption and constraint log is dynamic and all activities and outcomes associated with the incorrect assumptions may need to be adjusted in the future.

Assumptions that are incorrect may lead to risks or issues. When this is the case, under the status column you would indicate that the assumption has been transferred to a risk or issue log.

References

1. Edmonds, B. (2006). What is complexity? In: *The Evolution of Complexity* (ed. F. Heylighen and D. Aerts). Dordrecht: Kluwer.
2. Dionisio, C.S. (2023). *A Project Manager's Book of Templates*. Hoboken, NJ: John Wiley and Sons.
3. Nieto-Rodriguez, A. (2022). *The Project Revolution*. London, UK: LID Publishing.
4. Dionisio, C.S. (2022). *A Project Manager's Book of Templates*. Hoboken, NJ: John Wiley and Sons.

PART

III

Project Planning Tools

4

STAKEHOLDER ENGAGEMENT

stakeholder is commonly defined as person who has a vested interest in the outcome of a project. More importantly, for a project manager, a stakeholder is anyone who can influence, either positively or negatively, the outcome of their project. This includes people and groups of people both inside and outside the organization. As project managers we engage with stakeholders to manage communication, conflict, politics, and other aspects of stakeholder interaction to ensure a positive outcome.

Accountability for success relies on the abilities of the project manager even though we have limited positional power within the organization. Building strong relationships and engaging with stakeholders successfully are a key aspect of successful project delivery.

Stakeholders are many and varied on a project and come to the table with a variety of expectations, opinions, perceptions, priorities, fears, and personal agendas that many times are in conflict with one another.[1] The challenge is to find a way to *efficiently* manage this cast of characters in a way that does not become all consuming. This includes identifying highly influential stakeholders and then developing a stakeholder strategy that balances their expectations with the realities of the project.

Identifying stakeholders should start at the very beginning of the project, sometimes even before project origination documents are complete. All activities associated with stakeholder engagement occur throughout the project.

The tools presented in this chapter are designed to help project managers develop an effective stakeholder engagement strategy to positively influence the attitudes and behaviors of the stakeholders associated with their project. Note the use of the term stakeholder engagement. Anyone who has managed a project knows that you can't manage stakeholders. At best, you can engage them effectively.

Project Management Toolbox: Tools and Techniques for the Practicing Project Manager,
Third Edition. Cynthia Snyder Dionisio, and Russ J. Martinelli.
© 2025 John Wiley & Sons, Inc. Published 2025 by John Wiley & Sons, Inc.

STAKEHOLDER REGISTER

A Stakeholder Register is the first piece of documentation used when identifying and analyzing project stakeholders. The register is used to document relevant stakeholders along with their key information. Even though every project manager can benefit from using a Stakeholder Register, it is most valuable in project situations that have a larger number of stakeholders and where their opinions and viewpoints about the project are not well aligned.

The Stakeholder Register is started at the beginning of the project and is continually updated with additional and/or refined information.

Developing a Stakeholder Register

The first step in developing a stakeholder register is to identify all stakeholders. For smaller internal projects, this is relatively easy. You would probably consider the team, the sponsor, the customer, the end user, and potentially senior management and re-source managers. However, as projects get larger, the stakeholder pool becomes more diverse, the relationships among stakeholders are more complex, and therefore the job of engaging stakeholders effectively becomes more challenging. Table 4.1 identifies a number of stakeholder sources to consider.

Table 4.1: Sample Stakeholder Sources

Internal Stakeholder Sources

☑ Project Team	☑ Project Sponsor
☑ Organizational top managers	☑ Functional (department) managers
☑ Business strategists	☑ Governance bodies
	☑ Information Technology group
☑ Internal clients	☑ Customer Service group
☑ Human Resources group	☑ Finance group
☑ Legal group	☑ Manufacturing group
☑ Procurement group	☑ Quality Assurance group
☑ Sales and Marketing group	☑ Operations group

External Stakeholder Sources

☑ Clients or customers	☑ Competitors
☑ Community organizations	☑ Vendors
☑ Regulators	☑ Trade unions
☑ Media	☑ Lobbyists
☑ Users	☑ Venture capitalists

Table 4.2: Sample Stakeholder Register

Name	Role	Contact Information	Requirements	Expectations
Sue Williams	Sponsor	Email/phone		
Ajit Verjami	Software Lead	Email/phone		
Steven Cross	Client	Email/phone		
Danielle Carvalho	Subject Expert	Email/phone		
Anna Tamara	Subject Expert	Email/phone		

In addition to a list of stakeholders, the stakeholder register should reflect the needs of the project. Common content in the register includes:

- Identification information such as contact information, department, location, and role on the project.
- Assessment information like stakeholder requirements and expectations. For smaller projects this is usually sufficient.
- Classification information can be used for large and complex projects and is dependent on the project. You may want to include a stakeholder's relative degree of power or influence on your project, whether or not they support your project, if they are internal or external to the project, and how interested they are in your project.

Table 4.2 shows a generic stakeholder register prior to conducting any stakeholder analysis.

Using the Stakeholder Register

The primary use of the Stakeholder Register is to help the project manager structure the various stakeholders into some form of logical order. This is particularly useful for projects with a large number of stakeholders.

The register will be updated with the information gained from performing a Stakeholder Analysis. The Stakeholder Register is a very dynamic tool, constantly changing, and for that reason it must be reviewed and updated on a regular basis. Project stakeholders transition in and out, as do some of their viewpoints and opinions about the project.

STAKEHOLDER ANALYSIS

Project stakeholders can be many, dispersed, and varied in their viewpoints and characteristics. It is important to do a good job in identifying your stakeholders, but stakeholder identification by itself has limited value for project managers. Developing a deeper understanding of the project stakeholder's interests, opinions, and viewpoints is the necessary next step in the stakeholder engagement process. This step is commonly referred to as stakeholder analysis. The results from the stakeholder analysis are used to further elaborate the Stakeholder Register.

Effective stakeholder engagement is supported by thorough stakeholder analysis activities that allow a project manager to develop a strategy to accomplish the following:

- Identify strategic interests that the various stakeholders have in the project to negotiate a common interest
- Develop plans and tactics to effectively negotiate competing goals and interests between stakeholders
- Secure active support for the project
- Devise activities to either neutralize or prevent the negative actions of those who do not support the project
- Allocate resources to engage with key stakeholders.

Many stakeholder analysis activities are about prioritizing the project stakeholders. A small subset of your stakeholders, commonly called your key or primary stakeholders, possess a significant amount of organizational influence to either advance your project or block its progress. Either way, these stakeholders can be identified through a prioritization process.

To avoid becoming overwhelmed with data, it is a good idea to figure out what information is needed to effectively engage with stakeholders. Then the project manager should maintain focus on the primary stakeholders to prevent the job of stakeholder engagement from being all-consuming to the detriment of other critical aspects of managing a project.

There are a number of effective tools that project managers use to analyze their stakeholders and ensure they are focused on the most critical stakeholders. A robust stakeholder analysis helps project managers develop a stakeholder strategy to gain maximum stakeholder alignment and to effectively maneuver the political landscape surrounding the project.

Conducting Stakeholder Analysis

Stakeholder Analysis begins during the earliest phase of a project, and continues as stakeholders and their thoughts about the project change. Introductory meetings help to understand each stakeholder's understanding, interpretation, and opinion of the project. Additionally, the meeting can be used to collect information on each stakeholder. Some of the qualitative data you can gather to analyze stakeholders includes:

- Requirements
- Expectations
- Reservations
- What they provide to the project (resources, funding, time).

Table 4.3 shows an example of how this information would be incorporated into the Stakeholder Register.

The qualitative assessment should provide enough information so that the primary stakeholders can be identified. These are the stakeholders who have the most impact on a project, particularly those who can impact critical decisions or resource assignments. Remember to include stakeholders who have the ability to negatively impact the project.

The number of primary stakeholders will typically relate to the size of a project. Larger projects normally require broader organizational and partnership involvement, and subsequently more stakeholders will be involved.

Once the primary stakeholders are identified there are several variables you can evaluate, such as:

- Influence
- Support
- Power
- Interest.

A common means of evaluating stakeholders is to create a matrix with two of these four variables. We will demonstrate this using Influence and Support.

Table 4.3: Example of Qualitative Stakeholder Analysis

Name	Role	Requirements	Expectations	Reservations	Provides to Project
Sue Williams	Sponsor	Deliver business value on time and budget	Project meets all business and execution goals	Ability to develop the new capability	Direction and decisions
Ajit Verjami	Department Manager	Requirements, features, and functions desired	No expectations for this project	Believe another project provides a better solution	Resources
Steven Cross	Client	Timely delivery of all functions	Project completed under budget	Timeline is very aggressive	Funding
Danielle Carvalho	Subject Expert	Clear statement of work to support the project	Project will be run efficiently	Already committed to two other projects	Time and expertise
Anna Tamara	Subject Expert	Statement of Work to develop a contract	Clear documentation of expected vendor deliveries	She is overwhelmed with other work	Contracts and legal advice
...	...	...	...	...	...

Evaluate the Level of Stakeholder Influence

For each primary stakeholder, evaluate the level of influence you believe they have on the outcome of the project. Focus on those individuals who can affect critical project decisions (such as phase gate approvals), those who are providing resources to the project, those who control all or portions of the project budget, and those who have expressed reservations about the project.

Plot the stakeholders with respect to their level of influence on the Stakeholder Evaluation Matrix by assessing whether you believe a stakeholder has a high, medium, or low level of influence on the project. Do this by asking the question "To what degree can the stakeholder exert influence to change the course of the project?"

The Difference Between Power and Influence

There is a difference between power and influence. Often, however, even among some of the most seasoned managers, power and influence gets confused. The following characterizes and distinguishes the two.

Power	Influence
Positional	Personal
Formal	Informal
Assumed	Earned
Control	Consent
Engagement	Leadership
Command	Request
Dictate	Dialogue

Power is an ability someone has to direct the behaviors and actions of others. Similar, but different is influence, which is a personal capacity that one has to affect the behavior of someone.

Evaluate the Level of Support

To evaluate the level of support you are assessing a stakeholder's attitude toward a project. Your evaluation is concentrated on trying to determine if a stakeholder is supportive, non-supportive, or indifferent toward your project. It is far better to understand if your stakeholders agree with the project goals that you have identified and documented early in the project cycle instead of waiting until it is too late to adjust the goals or work with stakeholders to get alignment. The project goals in turn affect the scope of work to be performed and the outcomes of that work.

Unless you have a very small number of project stakeholders, it is unrealistic to believe that there will not be conflicting opinions and interests between the stakeholders. This is common, and for this reason it needs to be uncovered and understood in order to develop a strategy and tactics to broker this conflict of interest. Left unchecked, conflicting interests and opinions are a leading cause for project failures.[2]

The Stakeholder Analysis can be used to begin determining who your supporters and non-supporters are. Specifically, pay close attention to the information that is contained in the "reservations" portion of the table. Typically, non-supporters will bring their reservations to light during the discovery conversations with the project manager. Normally an additional level of analysis is needed to determine if stakeholders who are not advocates may in fact become blockers to project success. However, at this stage project managers should be able to separate their advocates from the non-supporters.

Plot the Stakeholders on the Matrix

Next you plot each stakeholder on a matrix based upon your assessed level of influence and support to the project, as shown in Figure 4.1. This provides a graphical representation of their stakeholders.

The stakeholders in quadrant A possess high influence and low, or indifferent, support of the project. These stakeholders constitute the highest risk to the project and will require significant focus and attention.

The stakeholders in quadrant B possess high influence and positive support for the project. These individuals or groups are potential project champions. Work will be required to ensure the continued support of these stakeholders. These individuals or groups may be instrumental in helping to move other stakeholders, especially those in quadrant A, to more favorable positions on the matrix.

The stakeholders in quadrant C are generally non-supportive of the project but possess little influence within the organization to affect the project outcome. These stakeholders cannot be ignored however, as organizational and political landscapes are

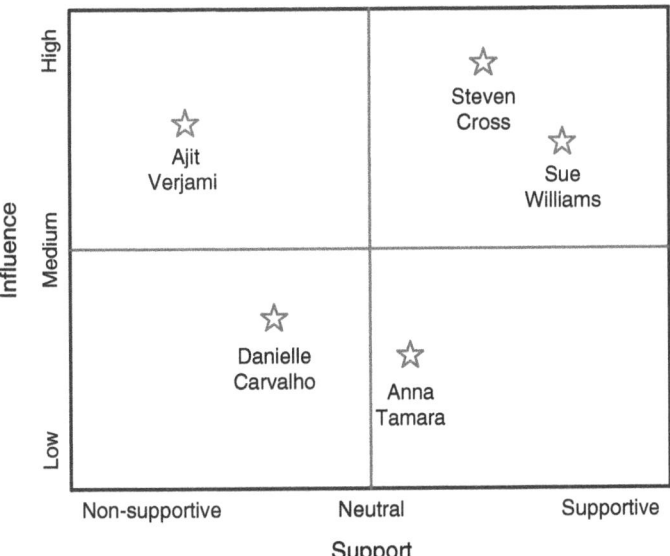

Figure 4.1: Example of a Simple Stakeholder Matrix

dynamic. It is not uncommon for a stakeholder originally placed in quadrant C to gain influence (through a promotion to a position of power for instance) throughout the course of a project. If they remain non-supportive once they gain additional influence, an additional risk to the project is introduced.

The stakeholders in quadrant D are supportive of the project, but again possess little influence over the project outcomes at the current time. Use these stakeholders as part of your project coalition of support by keeping them informed and engaged. Often these stakeholders possess specific knowledge and viewpoints that can be leveraged to support specific decisions or actions you are trying to influence.

Plot Stakeholder Movement

Using the information gained by segmenting the evaluation matrix into a grid, physically draw the various stakeholder movements that will be necessary to secure positive support for the project. This provides a visual representation of the *future state* of the project stakeholder landscape. This future state provides clarity and focus for the development of your stakeholder engagement actions.

Be realistic when plotting your planned stakeholder movement. For instance, it is not realistic to believe you can move a highly influential stakeholder with strong aversion to your project to a position of strong support. In reality, the best scenario may be that you will be able to move this stakeholder to a neutral position. This in itself is a wonderful risk mitigation strategy.

Keep in mind that sharing the information shown in Figure 4.2 may be a career-limiting move. Detailed stakeholder analysis information should be considered confidential and

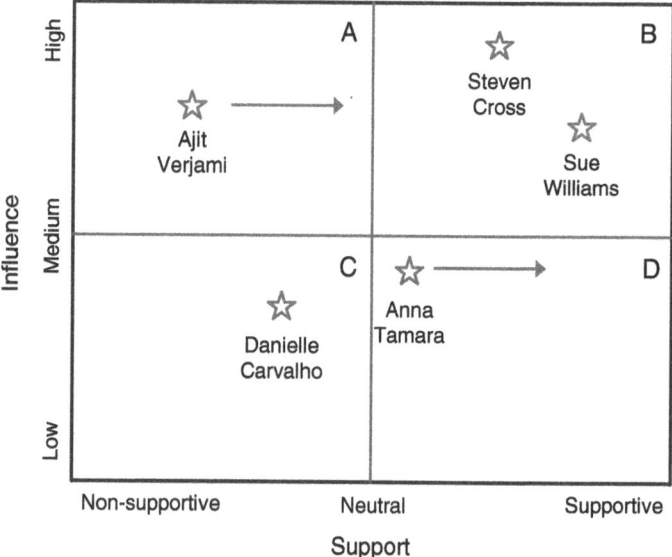

Figure 4.2: Example of a Desired Future State Stakeholder Matrix

tightly controlled by the project manager. Why do we say this? Because true and honest stakeholder analysis brings to the surface the realities of corporate politics and unique biases that have to be managed properly; usually that means actively, but delicately.

Variations

You can build out the information in the stakeholder analysis matrix by providing more detail. Figure 4.3 shows a more robust analysis that incorporates the stakeholder, and their role on the team. We will continue using information on influence and support, but rather than plotting the information on a matrix, it is shown in tabular format with high (H), medium (M), and low (L) influence, and support shown with arrows pointed up or down, and a bar to indicate neutral support. The type of contributions each stakeholder provides to the project are shown with an X to indicate the type of contribution.

Keep in mind that this is only an example. It is up to each project manager to determine what information they need in order to meet their needs.

Another variation is a tool called the Powergram (see Figure 4.4). The Powergram focuses exclusively on the power dynamics between project stakeholders.[3] It is especially useful in project situations where achieving alignment between stakeholders is a challenge caused primarily by differences in opinion or historical tensions between individuals.

The Powergram displays the power structure between key stakeholders which can provide insights into where stakeholder tension exists and how other stakeholders can be used to positively influence and change the power dynamics.

Stakeholder Name	Relationship to Project	Level of Influence	Support to Project	Resources					
				People	Money	Material	Facilities	Knowledge	Decisions
Sue Williams	Sponsor	H	⇑		X			X	X
Ajit Verjami	Functional Manager	H	⇓	X		X			X
Steven Cross	Software Vendor	L	▭	X			X	X	
Lynda Donovan	Core Team Member	M	⇑					X	

Influence Key: L = Little or no influence M = Some influence H = Considerable influence

Support Key: ⇑ Positive support ▭ Neutral support ⇓ Negative support

Figure 4.3: Example of a Robust Stakeholder Matrix

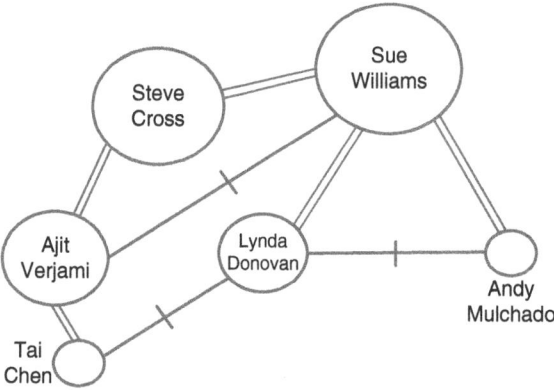

Figure 4.4: Example Powergram

Developing a set of power relationship rules is necessary to construct an effective Powergram. We suggest the use of the following rules:

1. The size of a circle denotes the amount of organizational power a stakeholder possesses. The larger the circle, the more power they possess.
2. A double line represents a *positive* relationship between two stakeholders.
3. A single line with a strike through it represents a *negative* relationship between two stakeholders.
4. The shorter the line connecting two stakeholders, the stronger the relationship between them.

The following power dynamics can be derived from the Powergram. Sue Williams (Project Sponsor) and Steve Cross (Client) are the most powerful stakeholders, with Ajit Verjami (Department Manager) relatively powerful as well. The Powergram also shows that there is a strained relationship between Williams and Verjami, but the positive relationship between Williams and Cross, and between Cross and Verjami, can be leveraged if needed. Finally, Lynda Donovan (Network Administrator) has negative relationships with both Chen and Mulchao. As a result, the positive relationships between Williams and both Mulchao and Donovan may be needed to gain alignment between these two project stakeholders.

The Powergram is a very simple tool to create and does not require a high degree of precision and accuracy to be useful in helping project managers to make sense of the power dynamics surrounding their project.

COMMUNICATION MATRIX

Once you have identified and analyzed your stakeholders you have a much better understanding of how to engage with them effectively. Communication with stakeholders is a key ingredient in establishing rapport, building relationships, and creating trust.

Table 4.4: Communication Methods			
Written	**Verbal**	**Electronic**	**Media**
Reports	Presentations	Email	Videos
Technical papers	F2F Meetings	Forums and Blogs	Audio
Project planning documents	Phone calls	Social media	Tutorials
Press releases	Briefings	AI generated notifications	Video conference

Communication can take place in written form, verbally, electronically, or through media. It can be formal or informal, internal or external. Table 4.4 gives a few examples of different types of communication methods. This is by no means a complete list.

Many of these methods are, by their nature, combined, such as verbal presentation with a supporting slide deck. Another way of categorizing communication methods is by how people receive the information. Common categories for receiving information include push, pull, and interactive communication.

- Push communication entails sending information directly to stakeholders. Examples include email, memos, and voice mail.
- Pull communication requires that stakeholders actively seek out the information. Examples include internet searches, internal lessons learned databases, and knowledge repositories. Pull communication is generally used when there is a large stakeholder group. For example, if you are doing a public works project you would have a high-level schedule, key project documents, information on road closures, and so forth available for the public to access as needed.
- Interactive communication is between two or more stakeholders. This includes face-to-face or video team meetings, phone calls, and presentations.

Before developing a communication matrix, you will want to consider any special circumstances that will influence how you communicate with stakeholders. In particular, pay attention to technology, cross-cultural communication, the political environment, and confidential information. So much of project work gets done via technology, and the advent of cloud-based computing makes it more likely that stakeholders will have access to the information they need. However, it still bears consideration of whether everyone has access to the technology they need to send and receive information as needed.

Cross-cultural communication can refer to age ranges, nationality, or even professions. Young professionals tend to be more comfortable sending and receiving information electronically, whereas baby boomers tend to prefer more interactive communication.

Having political awareness when developing and using a communication matrix means understanding the power structure within an organization. This includes knowing who has the authority to make decisions, as well as who has informal power and influence. The information on the stakeholder matrix can help to develop an effective strategy for politically astute communication.

Finally, due consideration should be paid to the potential sensitivity of information. Some information is appropriate for external communication, some should be considered business sensitive and only be released to internal stakeholders, and still other information may be considered secret or highly sensitive for limited release only.

Developing a Communication Matrix

A communication matrix answers the following questions:

- Who needs information?
- What information do they need?
- When do they need it?
- In what format should it be delivered?
- Who is accountable for delivering the information?

For most projects that are not large or complex, this information is sufficient. It is most commonly displayed as a table, such as that shown in Table 4.5

Your stakeholder matrix may be best suited by identifying specific individuals, or identifying groups of stakeholders, such as where people are on the stakeholder matrix, or vendors, team members, and senior management. Once you have identified the basic information, you will want to tailor it to meet the specific needs of your stakeholders. For example, if you have an international team, you will want to designate a primary time zone and a primary language so people communicate consistently.

You may need to expand your matrix to include information specific to the project, such as a glossary of terms, a list of acronyms and abbreviations, or a flowchart of communication processes.

Using a Communication Matrix

The communication matrix is a key component of planning for effective stakeholder engagement. It supports you in organizing and optimizing stakeholder engagement and reduces the chances of misunderstandings. Once it is established, it will need to be maintained as new stakeholders arrive, others leave, and as you find ways to improve and fine tune the information in the matrix.

Table 4.5: Communication Matrix

Stakeholder	Information	Method	Timing/Frequency	Sender
Sponsor	Status reports, risk report, significant change requests	Email	Monthly	Project Manager
Vendor	Statement of work, schedule	Meeting	Start up, then monthly	Project Manager
Project manager	Progress reports, risks and issues	Email	Weekly	Vendor 1

You can improve the effectiveness of communication by remembering a few key tips for effective communication:

1. *Be clear on the purpose of each communication.* What result do you want from the communication?
2. *Structure for clarity.* Whether you are delivering information in person or electronically, verbally or in writing, take time to structure the information to provide the most clarity and the least opportunity for misunderstanding.
3. *Be concise.* Eliminate any superfluous words and information.
4. Use graphics, charts, and images to enhance the message and reduce reliance on words.
5. Remain aware of cultural differences so the information is respectful and impactful.
6. For written communication, make sure your spelling and grammar are correct.
7. For interactive communication, listen closely, check for understanding, and summarize key points.

STAKEHOLDER ENGAGEMENT PLAN

The Stakeholder Engagement Plan pulls together the information from the stakeholder register, stakeholder analysis, and communication matrix. It establishes the framework the project team will use to identify, categorize, and analyze key project stakeholders in order to develop and implement an effective stakeholder strategy.

Developing a Stakeholder Engagement Plan in the early stages of a project can help to align your key stakeholders to the goals of the project and to turn the key stakeholders into advocates for project success. Having a stakeholder plan in your PM ToolBox gives you an opportunity to proactively influence stakeholders so when issues arise, the most influential stakeholders are positioned to assist instead of hinder.

Developing a Stakeholder Engagement Plan

A Stakeholder Engagement Plan is an effective tool to assist a project manager in creating a focused stakeholder strategy by identifying the correct set of stakeholders, aligning the stakeholders to the goals of the project, and building good personal and professional relationships with key stakeholders. It ensures that the effort spent in engaging stakeholders is not happening in a haphazard manner. Rather, that the efforts are focused on the right set of stakeholders and that they are working toward achieving the desired outcomes.

Even though every project manager can benefit from using a Stakeholder Engagement Plan, it brings the most value to projects with a larger number of stakeholders, especially when their opinions and viewpoints about the project are not completely aligned. In these project situations, the information documented the Stakeholder Engagement Plan is a wonderful tool to assist project managers in creating structure out of chaos.

Stakeholder Engagement Approach

The Stakeholder Engagement Plan is a document that is developed as part of the broader project management plan and establishes the framework for how a project manager will perform their stakeholder engagement activities. Included in this plan is a general description of the approach used to identify, analyze, and influence project stakeholders.[4] It should include information about the following:

- *Stakeholder engagement process.* The plan should identify and describe how engagement of stakeholders will be performed on the project.
- *Roles and responsibilities.* The overall responsibility for stakeholder engagement lies with the project manager, but he or she must leverage members of the project team and assign responsibility to them as appropriate. The plan should identify members of the team who have responsibility for engaging with stakeholders as well as which stakeholders they are responsible for establishing and maintaining a relationship with.
- *Tools.* The plan should describe the various tools that will be used to identify stakeholders, to analyze the stakeholders' interest, attitudes, power, and influence in order to develop and implement the stakeholder strategy.

Identifying Stakeholders

Effective stakeholder engagement begins with identification of all stakeholders associated with a project. The Stakeholder Engagement Plan should discuss how project stakeholders will be identified. It is important that the list of stakeholders should be comprehensive in order to cast a wide net over all players who may have a vested interest in the outcome of the project.

Tools such as a stakeholder register are effective in helping a project manager identify the various stakeholders. The stakeholder list should include internal stakeholders such as top managers, project governance board members, department or functional managers, support personnel (accounting, quality, human resources), and the project team. External stakeholders should also be listed and may include contractors, vendors, regulation bodies, service providers, and others. The objective of stakeholder identification is to include anyone who *might* have an influence on the outcome of the project.

Stakeholder identification includes the categorization of stakeholders into the logical groups that they belong to. Such categories may include senior sponsors, executive decision makers, team members, and resource providers to name a few. It is important to realize that some stakeholders may belong to multiple groups. The intent of stakeholder categorization is to bring structure to the stakeholder list based on common interests in the project.

Analyzing Stakeholders

The Stakeholder Engagement Plan should describe how the project team will analyze its list of identified stakeholders. Stakeholder analysis activities involve determining what type of influence each stakeholder has on the project, such as decision power, control of resources, or possession of critical knowledge, and their level of support to the project. In other words, would the stakeholder prefer the project to succeed, not to succeed, or is he or she indifferent about the outcome of the project?

The stakeholder analysis process is geared toward identifying the subset of stakeholders who should be deemed as *primary stakeholders* along with the reasoning for why they are deemed as key to project success. Primary stakeholders are often those who have the most influence over the outcome of a project (positive or negative) or those who may be the most affected by the project.

Being able to determine the primary stakeholders is critical to the project manager as he or she will not have time to engage with all stakeholders. To effectively build relationships with the right stakeholders, the project manager must flush out who the primary stakeholders are, understand what level and type of influence they have, and determine how they feel about the project. These stakeholders will therefore require more communication and engagement to solicit and maintain their support.

Creating a Stakeholder Strategy

Stakeholder analysis is about sense-making. This means understanding the significance of the information gained about the various project stakeholders. Significance of the information is then used to develop a strategy for engaging and managing the *right* set of stakeholders. The right stakeholders are those who have power and influence to affect the outcome of the project.

Most of the literature on stakeholder engagement immediately classifies the primary stakeholders as those with both high power and strong support for the project. This is significantly inaccurate because it leaves out the most potentially dangerous stakeholders—those with high power and no support for the project. These people also need to be considered primary stakeholders. Figure 4.5 helps to illustrate why. Stakeholders with high power and negative allegiance fall in the category of significant engagement required.

If a project manager uses the power/support grid to map their stakeholders, a stakeholder strategy will begin to emerge. The strategy should consist of a communication and action plan for each of the primary stakeholders. It should also keep the project advocates engaged, describe how they can be used to influence others, and plan how to win over or neutralize the stakeholders who are not current advocates.

Using the Stakeholder Engagement Plan

Projects often involve stakeholders who rarely come to the table completely aligned in full support of a project. Because of this, a stakeholder strategy needs to be developed and followed to ensure stakeholders are engaged effectively to set the project up for success.

The Stakeholder Engagement Plan is normally created during the earliest stages of a project, either during project initiation or project planning. Many times, it is incorporated into a larger project plan that covers all aspects of planning a project effectively.

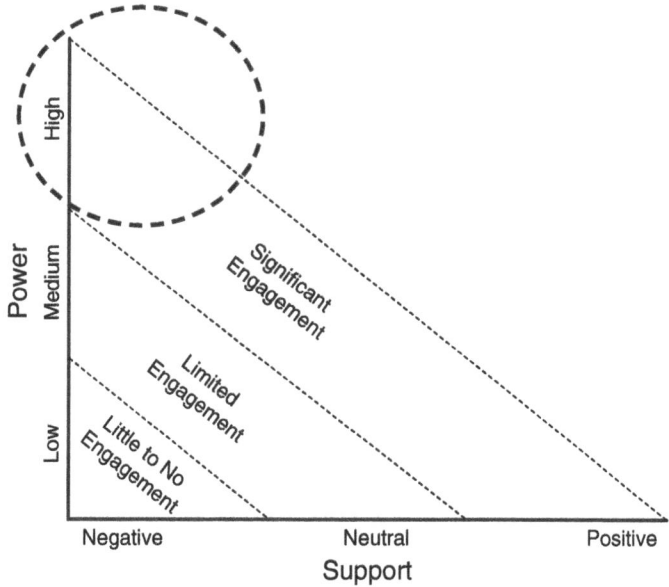

Figure 4.5: The Power/Support Grid

For smaller projects, only a few hours may be required to conduct a planning session and develop a plan. This time proportionately rises as projects get bigger and more stakeholders become involved. Tens of hours may be necessary to devise a quality Stakeholder Engagement Plan for a team in charge of a large and complex project.

Even though it is created early, the plan should be viewed and treated as a dynamic tool. This is due to the fact that project stakeholders are often transient, entering and exiting throughout the project cycle. As turnover in stakeholders occurs, it is likely that the Stakeholder Engagement Plan will need to be modified.

CHOOSING YOUR STAKEHOLDER MANAGEMENT TOOLS

The tools presented in this chapter are designed for various project stakeholder management activities and situations. Matching the tools to their most appropriate usage is sometimes a bit confusing. To help in this effort, the following table lists various stakeholder management situations and identifies which tools are geared for each situation. Consider this Table 4.6 as a starting point, and create your own custom project situation analysis and tools of choice to fit your particular project management style.

Table 4.6: The Power/Support Grid

Situation	Stakeholder Engagement Plan	Stakeholder Register	Stakeholder Analysis	Communication Matrix
Identify project stakeholders		√		
Determine stakeholder interest in the project			√	
Determine how stakeholders will be impacted by the project			√	
Identify primary stakeholders			√	√
Assess stakeholder characteristics				√
Establishes project stakeholder management methodology	√			
Used to deliver a stakeholder engagement strategy			√	√

References

1. Bourne, L. (2016). *Stakeholder Relationship Management*. Routledge.

2. IBM *IBM Systems Magazine*. www.ibmsystemsmag.com/power/Systems-Management/ Workload-Management/project_pitfalls/?page=2. Accessed April 2015.

3. Andler, N. (2016). *Tools for Project Management, Workshops and Consulting: A Must-Have Compendium of Essential Tools and Techniques*, 3rd ed. Erlangen, Germany: Publicus Publishing.

4. Singh, H. (2014). *Mastering Project Human Resource Management: Effectively Organize and Communicate with all Project Stakeholders*. New York, NY: Pearson FT Press.

5
REQUIREMENTS MANAGEMENT

The successful delivery of any project hinges on a clear, well-defined scope. However, the project scope is directly shaped and informed by the requirements gathered from stakeholders. It is essential that the requirements management process be tightly integrated with scope management to ensure alignment between what the project aims to deliver and what stakeholders expect.

This chapter will explore the key elements of effective requirements management, from systematically eliciting and specifying requirements to developing a comprehensive plan for identifying, analyzing, and tracking requirements. Crucial to this process is establishing robust traceability between the defined requirements and the approved project scope, including the work breakdown structure (WBS). By taking an integrated approach to requirements and scope, project teams can enhance their ability to meet stakeholder expectations and deliver a successful outcome.

While all projects have requirements, digital projects, new product development, and system engineering projects rely heavily on them. Therefore, many of the examples in this chapter will refer to this type of project.

The area of requirements uses language in very specific ways. The section on Requirements Language has specific terminology and definitions for requirements.

Requirements Language

Requirement. A capability that must be present, or a need that must be met to achieve the project objectives.
Elicitation. A structured approach to draw out requirements.
Specification. A detailed document that precisely describes the features, functions, and constraints of a product or system.

(continued)

Project Management Toolbox: Tools and Techniques for the Practicing Project Manager,
Third Edition. Cynthia Snyder Dionisio, and Russ J. Martinelli.
© 2025 John Wiley & Sons, Inc. Published 2025 by John Wiley & Sons, Inc.

Traceability. Tracking requirements by establishing linkages, such as to deliverables, stakeholders, or other requirements.

Verification. Determining if a requirement or deliverable fulfills acceptance criteria, or is technically correct.

Validation. Ensuring that a requirement of deliverable meets the needs of the customer or other stakeholder.

REQUIREMENTS ELICITATION PLAN

When talking about requirements the experts use the term "elicitation" when referring to uncovering requirements. Rather than using the term "gather" or "collect" which infer that requirements are easily identified and documented, the term "elicit" emphasizes the need to uncover or draw out requirements. Sometimes requirements are easy to identify and can be "collected," but other times it requires more skill and analysis to uncover hard-to-identify requirements.

An elicitation plan ensures that requirements are uncovered from all impacted stakeholders, including those requirements that are implicit or challenging to identify. Taking the time to consider the best way to elicit requirements from each stakeholder or each group of stakeholders promotes a structured approach for ensuring a thorough process.

By taking a proactive, planned approach to requirements elicitation, the project team can lay the groundwork for a successful requirements management process and, ultimately, the delivery of a solution that meets stakeholder needs.

Developing a Requirements Elicitation Plan

Often a business analyst will manage the entire elicitation process. However, they may rely on team members with subject matter expertise to elicit technical requirements or specialty requirements, such as testing or security requirements.

The first step in creating an elicitation plan is to identify the key stakeholders who will be involved in requirements elicitation. This typically includes the project sponsor, subject matter experts, end users, and any other parties who will be impacted by or have a vested interest in the project deliverables. Chapter 4 has information on identifying and analyzing stakeholders.

The resource(s) responsible for eliciting requirements will be involved in determining the appropriate elicitation technique for each stakeholder. Two common categories of eliciting requirements are working with stakeholders and using Mockups.

Working with Stakeholders

Keep in mind that many times stakeholders don't know what their requirements are; therefore, it is useful to have several elicitation methods to help you uncover their needs. The most common way of eliciting requirements is by working directly with your stakeholders. There are multiple methods you can use to draw out or elicit requirements from stakeholders.

Interviews

One-on-one discussions with stakeholders is an easy way to help uncover what they want and need. It helps to ask both open-ended and closed-ended questions. For example, you may start the interview with open-ended questions such as: "What would you like this system to do?" or "What are the current problems you are having?" Depending on the responses you might start to narrow the questions down, such as: "Given that this system will be using information from multiple sources, which source should be prioritized?" or "How would you like the outputs displayed?" Using both open-ended and closed-ended questions helps generate responses that can be later documented as distinct requirements statements.

Focus Groups

A focus group is used to gather wants and needs from diverse groups of stakeholders. Focus groups are generally led by a skilled facilitator. You should prepare the facilitator with the types of questions you want, the interactions you are looking for, and so forth. Some focus groups include the option to watch and listen unseen, so you can get unbiased information. This technique allows you to capture a wide range of perspectives and identify common themes that can be translated into project requirements.

Brainstorming

Brainstorming sessions bring together a group of stakeholders to collaboratively generate ideas and identify potential requirements. These sessions are guided by a facilitator who encourages open, uncritical participation to foster creativity and build on participant inputs. Brainstorming creates an environment where stakeholders can freely share their thoughts without fear of judgment, leading to a broader set of requirements to consider.

Observation

Observing stakeholders as they perform their day-to-day activities can uncover implicit needs and requirements that stakeholders may not explicitly state. The project team can watch users interact with existing systems or processes, take note of pain points or workarounds, and gather insights that supplement the information gathered through other elicitation methods.

Observation provides a window into the real-world context in which the final solution will be used. It can also uncover information that stakeholders are unaware of or are not willing to share. One of the things to beware of is that people often behave differently when they are being observed, so you may not see a true representation of interaction.[1]

Mock-Ups

In addition to directly engaging with stakeholders, the elicitation process can also leverage various types of mock-ups and prototypes to help uncover requirements.

Prototypes

Developing preliminary versions or models of the product or service can be an effective way to solicit feedback from stakeholders. These prototypes, even if they are low fidelity, allow stakeholders to interact with a tangible representation of the solution. This can reveal requirements that may not have surfaced through interviews or focus groups alone. Stakeholders can provide input on the functionality, usability, and design of the prototype, guiding the project team toward the true underlying needs.

Storyboards

Graphical representations of user workflows and interactions, known as storyboards, can help stakeholders visualize how they might use the final product or service. Storyboards can be a picture, such as the storyboard for a revitalized town square in Figure 5.1, or they can show processes in a step-by-step fashion to bring the proposed solution to life. Stakeholders can provide feedback on whether the storyboarded interactions accurately reflect their expectations and requirements, allowing the project team to make adjustments early in the process.

Models

Physical or conceptual models of the product or service can also be leveraged during elicitation. These models may take the form of mockups, wireframes, or three-dimensional models on a small scale, like the one shown in Figure 5.2. They allow stakeholders to see and interact with representations of the future state. Much like prototypes, models

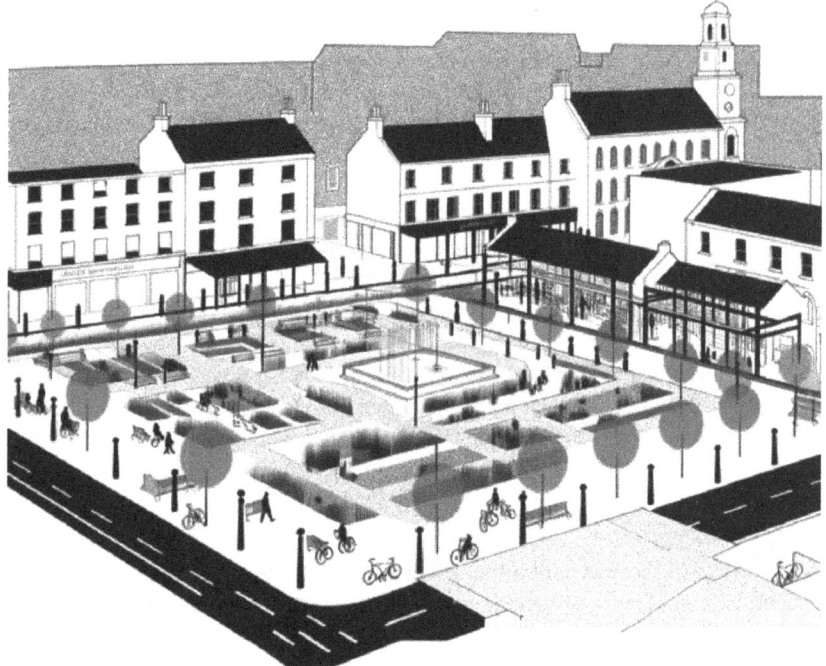

Figure 5.1: Town Square Storyboard
Source: Dall-E

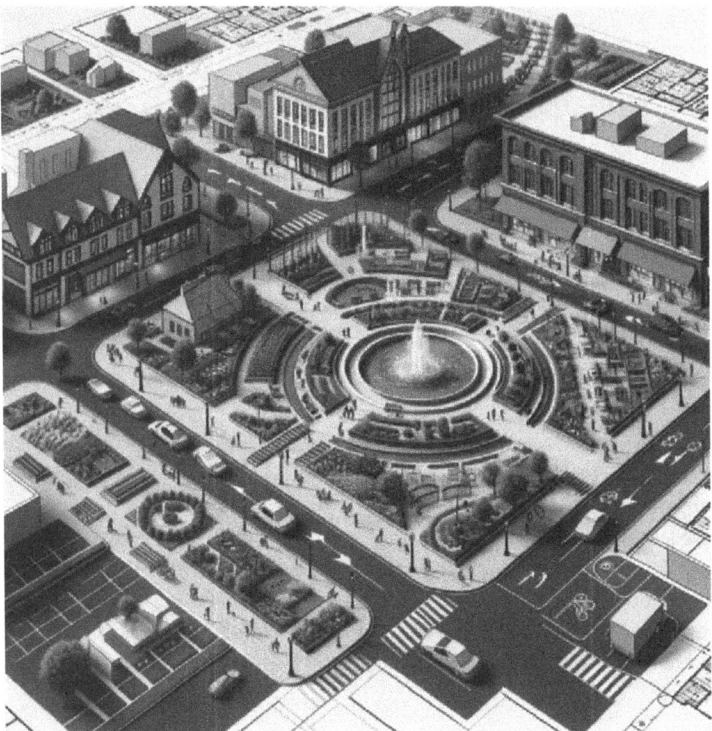

Figure 5.2: Town Square Model

enable stakeholders to provide input on the design, features, and overall functionality that should be included to meet their requirements.

Documenting the Requirements Elicitation Plan

Some stakeholder groups benefit from using multiple techniques for eliciting requirements. For example, end users can provide information via brainstorming, in a focus group, and via observation. Because requirements elicitation is often an iterative process, you may want to start with interviews or brainstorming, and then build prototypes of models to confirm understanding and elicit additional requirements.

In addition to the stakeholder and the requirements elicitation technique, it is useful to categorize the types of requirements. For example, you may categorize requirements as business requirements, user requirements, security requirements, and so forth. The output doesn't have to be elaborate; it can be a simple chart as shown in Table 5.1.

Using a Requirements Elicitation Plan

During the requirements elicitation process, the project team will execute the activities outlined in the elicitation plan. This involves scheduling interviews, facilitating focus

Table 5.1: Requirements Elicitation Plan			
Stakeholder	**Elicitation Technique**	**Category**	**Resource**
Sponsor	Interview	Business	Business Analyst
Internal customer	Interview Prototype	Functional	Business Analyst
End user	Brainstorming Focus groups Observation Story boards Models	Functional	Business Analyst Facilitator
Test lead	Interview	Technical	Quality Assurance

groups, or developing prototypes to present to stakeholders. The team should carefully document all feedback and insights gathered, as this information will feed directly into the requirements specification.

The elicitation plan serves as a roadmap to ensure a comprehensive and systematic approach to capturing requirements from all relevant stakeholders. The Requirements Elicitation Plan can be a standalone document, or it may be integrated into an overall Requirements Management Plan. As the project progresses, the plan may need to be updated to adapt to changing circumstances or newly identified stakeholder groups.

REQUIREMENTS MANAGEMENT PLAN

A Requirements Management Plan is a comprehensive document that outlines how requirements will be identified, analyzed, documented, and managed throughout the lifecycle of a project. It serves as a touchstone for stakeholders, guiding them on how the project's needs and expectations will be captured, prioritized, and translated into a deliverable solution.

For small to mid-sized projects, the Requirements Elicitation Plan may be sufficient for working with requirements. For larger, more complex projects, the Requirements Management Plan can be used to guide the development of a more detailed Requirements Elicitation Plan, Requirements Traceability Matrix, and Requirements Specification document.

Implementing a Requirements Management Plan enhances clarity and alignment among project stakeholders, ensuring that all parties have a shared understanding of the project's objectives and requirements. This alignment reduces the likelihood of scope creep and project delays, as it outlines clear processes for managing changes to requirements.

Developing a Requirements Management Plan

A Requirements Management Plan takes time and thought to craft. Before beginning you should review the Business Case (Chapter 2) for the high-level business requirements, the Project Charter (Chapter 3), the Scope Statement (Chapter 6), and the Stakeholder Register and Analysis (Chapter 4). These documents provide a summary of the project and stakeholders that you will use when determining the best approaches for eliciting, prioritizing, and managing requirements.

When developing the Requirements Management Plan it is wise to consider how scope planning will be conducted and maintain alignment. For example, as you consider how to elicit and prioritize requirements, think about how you will organize your WBS (Chapter 6). Because mismanaging requirements is one of the main reasons projects struggle or fail, consider how to structure change management to ensure any new requirements or changes to requirements are identified, evaluated, and approved (or not).

Requirements Categories

There are different categories of requirements. When working with a large project sorting on the various types helps prioritize, elicit, and organize your requirements. A common way of categorizing requirements is business requirements, user requirements, functional requirements, non-functional requirements, and technical requirements.

Business requirements are high-level statements of the goals, objectives, and needs of the organization as a whole. This type of requirement focuses on the business perspective of the system and is primarily concerned with ensuring that the system meets the strategic goals of the organization.

User requirements state the needs of a particular user class, including the tasks they must perform, the data they need to input, and the output they expect to receive. These requirements are often documented using narratives or user stories. They help ensure the system is user-friendly and meets the practical needs of its end-users.

Functional requirements define the specific behavior or functions of a system. They describe what the system should do in terms of tasks, services, or functions performed, particularly in response to specific inputs or in specific situations. These requirements might detail calculations, data processing, user interface interactions, and other features that fulfill the system's objectives.

Non-functional requirements specify the overall qualities or attributes of a system, such as performance, usability, reliability, and security. Unlike functional requirements that describe what the system does, non-functional requirements focus on how the system performs a particular function. They can include specific criteria like response times, data integrity, and the ability to handle concurrent users.

Technical requirements define the technical issues that must be considered to successfully implement a project. These can include specifications of hardware and software, system integrations, and configurations. Technical requirements also detail the technical standards, protocols, and methodologies that need to be adhered to during the project.

Different categories of requirements respond well to different elicitation techniques and help organize and prioritize the work.

Requirements Elicitation

If you are not using a Requirements Elicitation Plan, as mentioned previously, you should consider the best way to elicit requirements from various stakeholders and stakeholder groups. The most common way of eliciting requirements is by talking to stakeholders. This includes interviews, brainstorming, and focus groups. These techniques are often used as an initial way to understand user needs and expectations.

Sometimes the best way to uncover requirements is by observing the current situation to see what works and what doesn't work, such as job shadowing. Job shadowing helps to understand user behavior and workflow, identifies bottlenecks, work arounds, and inefficiencies. Observation also allows analysts to uncover unarticulated needs that can enhance functionality and streamline tasks.

Models, storyboards, and prototypes allow stakeholders to see simplified or scaled down versions of a process, product, or system. A visual representation, step through of a process, or the ability to actively engage with a deliverable often reveals additional requirements and is helpful in validating the requirements prior to baselining.

Prioritize Requirements

It is almost certain that you will have more requirements than your time and budget can accommodate. Another certainty is that there will be times when stakeholder requirements conflict with each other. In these situations you need a way to prioritize your requirements. Using a MoSCoW method, as shown in Figure 5.3, is an excellent option. MoSCoW stands for Must Have, Should Have, Could Have, and Won't Have. This method guides the development team and helps manage stakeholder expectations. A good place to start is ensuring the "must have" and "should have" requirements are included. Must haves include requirements for safety, legal and compliance. If time and funding allow, you can incorporate requirements from the "could have" column. The "won't have" column should align with information from exclusions noted in the Scope Statement (Chapter 6).

Another option is to plot requirements on a 2×2 grid. For example, you can evaluate the risk associated with the requirement on one axis and the value on the other, as shown in Figure 5.4. Those requirements with high-value and low-risk are the highest

Must Have	Should Have	Could Have	Won't Have

Figure 5.3: MoSCoW Analysis

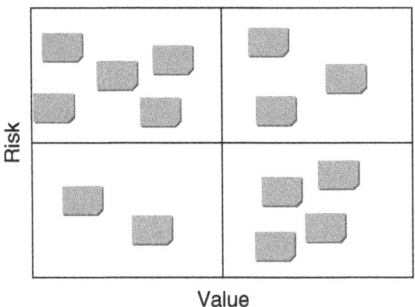

Figure 5.4: Risk-Value Matrix

priority requirements. The high-risk and high-value requirements should come next. Incorporate low-risk and low-value if time and cost allow. Those that are high risk and low value are probably not worth including.

Another option for a 2×2 matrix is to evaluate the difficulty to deliver the requirements and the value it provides.

Requirements Documentation

Define how requirements will be documented. The format of a requirements document may range from a simple spreadsheet to more elaborate forms containing detailed descriptions and attachments, or a requirements management system. For large or complex projects, you would develop a Requirements Specification. The Requirements Management Plan would describe the information in that would be incorporated in the Requirements Specification and the process for keeping it up to date.

Requirements Traceability

Tracing requirements is identifying, documenting, and following each requirement from inception to completion. This is often done with a requirements traceability matrix described further on in this chapter. There are several aspects of requirements that can be traced, for example:

- A requirement can be traced to the stakeholder that has the requirement.
- A requirement can be traced to the deliverable that fulfills the requirement.
- A requirement can be traced to the verification or validation method.

The Requirements Management Plan will specify what will be traced and how it will be traced.

Requirements Configuration Management

Managing requirements can quickly get out of control. Instituting configuration management for requirements helps reduce the likelihood of creeping and changing requirements.

Configuration management is the process of establishing and maintaining the physical attributes, requirements, design, and operational information for a system. Applying this practice for requirements management ensures that any changes or modifications to requirements are tracked, analyzed, and implemented systematically to maintain the integrity and traceability of the requirements and deliverables.

The Requirements Management Plan defines what will be under configuration control, how configuration management will interact with change management, how compliance with configuration management will be audited, and the authority levels associated with changes to requirements.

Verification and Validation

Requirements need to be validated to ensure they meet stakeholder needs, and verified to ensure they are implemented correctly. The four methods used to verify and validate requirements include inspection, testing, demonstration, and analysis.

Inspection involves reviewing documents, code, design diagrams, and other outputs to ensure compliance with requirements. Inspections are often conducted by peers or experts who analyze the artifacts to detect errors, inconsistencies, or deviations from the standards and requirements specified for the project.

Testing involves executing the feature, function, or system to ensure the deliverables functions as intended. This method checks the correctness, completeness, and quality of an output. Tests can be pass/fail or they can be quantitative measurements of how a deliverable, unit, or system performs.

Demonstration shows how a requirement is met based on the performance of a component or function under specified conditions. Demonstrations are typically less formal than testing and are used to prove capability and operational effectiveness to stakeholders without the stringent controls and environments associated with formal testing.

Analysis involves using mathematical models, simulations, or reasoned arguments to predict whether requirements are met. This method is used for verifying performance requirements or other characteristics that are difficult to test directly. Analysis can provide insights into system behavior under various scenarios, helping to confirm that requirements have been satisfied.

Large and complex projects may have a separate Verification and Validation Plan that is linked to the overall Requirements Management Plan.

The type of project and the needs of the stakeholders dictate the information in Requirements Management Plan. The information in Table 5.1 represents common information found in the plan.[2]

Using a Requirements Management Plan

A Requirements Management Plan is utilized throughout a project to ensure that the project stays on track, meets its objectives, and adheres to its scope. From the initial stages of a project, it provides a clear framework for eliciting, prioritizing, and managing requirements. The Requirements Management Plan provides a common understanding about the requirement process to project stakeholders.

As the project progresses, the plan serves as a living document that guides the requirements management process, ensuring that all requirements are documented, traceable, and maintained against the evolving project landscape. It also plays a role in change management, providing a structured approach for assessing the impact of changes and making informed decisions.

REQUIREMENTS SPECIFICATION

Requirements specification outlines the detailed capabilities, characteristics, and constraints that the project deliverables must meet in order to satisfy stakeholder needs. This specification forms the foundation for subsequent project planning, design, and verification activities.

After eliciting requirements, you have a set of wants and needs. Requirements specification transitions the raw set of requirements to a comprehensive set of documented requirements that are clear, comprehensive, and actionable. While a robust Requirements Specification is not necessary in every project, projects that involve digital outputs or system engineering projects will find this invaluable. The specification ensures that requirements are complete, consistent, unambiguous, agreed upon, and measurable.

Specifying Requirements

There are several steps to move from the initial set of requirements to the Requirements Specification. The following five steps provide a structured approach for the process that begins with organizing and categorizing to approval and baseline development.

Organize and Categorize Requirements

A long list of requirements with no organizing structure is at best challenging to work with, and for large projects, next to impossible. Therefore, the first step is to organize all the requirements into categories. A common way of organizing them is:

- Business requirements represent the goals, objectives, and outcomes of the system. These are often documented in the business case or charter. They are carried over to the Requirements Specification to ensure the project stays aligned with the intended outcomes. Examples include streamlining operations, improving customer satisfaction, or increasing market share.
- Functional Requirements that describe what the system or output should do. This includes processes, data handling, and functionality.
- Non-Functional Requirements that define how the system performs. This can include reliability, security, maintainability, and so forth.
- Constraints identify any restrictions. Restrictions can take the form of regulatory, legal, and safety constraints.

While not as common as a work breakdown structure (see Chapter 6), a requirements breakdown structure is a useful way to organize requirements. It typically starts with the high-level business requirements at the top level. The next level down may show major deliverables or functions. This is then decomposed into functional and non-functional requirements.

Refine Requirements

Prior to creating the Requirements Specification document, you need to make sure that each requirement is clear and detailed. Specifically, you want to ensure:

- *Requirements are clear and unambiguous.* There should only be one way to interpret a requirement—avoid ambiguous language. Each statement should only contain one requirement. Provide quantitative and measurable information where possible.
- *Requirements are appropriately detailed.* As necessary, include business rules, interface information, data flows, or other information that help provide context and understanding of the requirement.
- *Verification and validation.* Each requirement should have a way to verify it is correct, and a way for stakeholders to validate that it meets their needs. Examples include demonstration, testing, inspection, and black box testing for those requirements that are not observable.

Document Requirements

When you have organized, prioritized, and refined your requirements, you are ready to document them. Begin with an overview of the entire system or output you are creating. Describe the purpose, scope, and any relevant definitions and references for the output. Then provide a narrative description of an overview of user needs, assumptions, dependencies, and the operating environment.

The bulk of the specification comprises the requirements, a unique identifier for each requirement, and how each requirement will be verified. For large projects, this is usually done with a requirements management software application.

You can include appendices for a glossary, references, and any other information necessary to understand the requirements.

Validate Requirements

Before finalizing the Requirements Specification, you will want to validate it with relevant stakeholders. You may want to do this in a single meeting, or if the system is complex or there are many stakeholders, you may need a series of meetings. Based on stakeholder feedback you will update the document to reflect their wants and needs.

Baseline Requirements

The final step is baselining the requirements. Once you have stakeholder approval you will baseline the requirements and begin the process of configuration management and change management.

Using the Requirements Specification

As much as we would like it if requirements, once specified, never changed—that is not realistic. While the Requirements Specification is not a fluid document, you should expect updates. This can happen due to missing a requirement, a change in the market or the internal environment, or a change in stakeholders. Therefore, you will want to maintain a rigorous change management system (see Chapter 12). Any changes should include a description of the change, the name of the person who requested the change, a justification, and the impact of the change. Pay special attention to any trickle-down effects where a change in one requirement trickles down to other requirements. Without this type of document, changes quickly get out of control, people end up working on different versions of requirements, and unruly stakeholders add requirements without going through the proper channels.

It is useful to periodically review the Requirements Specification to ensure it remains consistent. Check to see if any conflicts between requirements have surfaced or if there are requirements that need to be further clarified or detailed.

As the project progresses the Specification document is used to verify and validate each requirement. The verification and validation methods are documented in the Specification, and testers use this information to ensure the requirements are met.

REQUIREMENTS TRACEABILITY MATRIX

Given that large complex projects may have thousands of requirements, how do you know which requirements will be affected by a change request? What if a deliverable is descoped? Which requirements will be eliminated? How do you ensure that a business requirement is represented with technical requirements? Trying to figure this out in your head is a recipe for a massive headache, if not a disaster!

A Requirements Traceability Matrix can help resolve all of these questions. A Requirements Traceability Matrix is used to track the various attributes or requirements throughout the project life cycle. It uses information from the Requirements Specification Document and traces how those requirements are addressed through other aspects of the project. For example, you can trace requirements back to business objectives or to the source of the requirement. You can also trace them forward to deliverables or verification methods.

The ability to trace and map requirements to multiple attributes provides a comprehensive view of requirements and their place in the project and the deliverables. It provides a succinct, up-to-date view of requirements and where they are in the process of being developed, tested, and accepted. Any proposed change can be evaluated to determine which requirements and deliverables are impacted by the change request.

Developing a Requirements Traceability Matrix

The first step in creating a Requirements Traceability Matrix is to figure out what you need to trace. Do you want to trace business requirements to technical requirements? If so, you would want to collect the information in Table 5.2.[2]

Table 5.2: Inter-Requirements Traceability Matrix

Document Element	Description
ID	Enter a unique requirement identifier
Business requirement	Document the condition or capability that must be met by the project or be present in the product, service, or result to satisfy the business needs
Priority	Prioritize the business requirement category, for example Level 1, Level 2, etc., or must have, should have, could have, or won't have
Source	Document the stakeholder who identified the business requirement
Technical requirement	Document the technical performance that must be met by the deliverable to satisfy a need or expectation of a stakeholder
Verification method	Describe the metric that is used to measure the satisfaction of the requirement

This type of traceability matrix may have a one to many relationship with business to technical requirements. It is useful to ensure all the technical aspects are present to meet the business requirements.

Another option is to trace the requirement to the deliverables that express that requirement. Table 5.3 shows a Requirement–Deliverable Traceability Matrix.

A Requirement–Deliverable Traceability Matrix may have a one to many relationship with technical requirements to deliverables, or vice versa. It is useful to ensure all the technical requirements are present in deliverables. Table 5.3 also shows how each requirement will be verified for correctness and validated for acceptance.

Table 5.3: Requirement–Deliverable Traceability Matrix

Document Element	Description
ID	Enter a unique requirement identifier
Requirement	Document the condition or capability that must be met by the project or be present in the product, service, or result to satisfy a need or expectation of a stakeholder
Source	The stakeholder that identified the requirement
Priority	Prioritize the requirement category, for example Level 1, Level 2, etc., or must have, should have, could have, or won't have
Category	Categorize the requirement. Categories can include functional, nonfunctional, maintainability, security, etc.
Deliverable	Identify the deliverable or WBS ID that is associated with the requirement
Verification	Describe the metric that is used to measure the satisfaction of the requirement
Validation	Describe the technique that will be used to validate that the requirement meets the stakeholder needs
Status	Where in the process the requirement is, for example, pending, in progress, or accepted

Variations

Traceability matrixes can be used in projects with evolutionary scope, such as those projects that use an Agile approach. For this type of project you may want to trace where in the process the requirement is. Using a task-board-type approach you can trace the requirement from backlog, to in progress, testing, and accepted.

Another way to trace requirements in adaptive environments is to show which iteration a requirement will be worked on, or which release it will be part of.

Using a Requirements Traceability Matrix

Unlike a Requirements Management Plan, a Requirements Traceability Matrix is meant to be a dynamic document. It can be updated as deliverables and requirements advance through the project life cycle. It can also be used as an audit tool to document how deliverables meet stakeholder requirements, or how business requirements are fulfilled by technical requirements.

References

1. Project Management Institute (2017). *The PMI Guide to Business Analysis* (ed. Project Management Institute, Inc). Newtown Square, PA.
2. Dionisio, C.S. (2023). *A Project Manager's Book of Templates*. Hoboken, N.J.: John Wiley and Sons.

6

SCOPE PLANNING

Project planning begins with identifying work that needs to be completed for a project to be successful. Scope planning is a collaborative process between the project manager, the project sponsor, and other key stakeholders who help shape the work to achieve the business goals driving the need for a project.

The purpose of scope planning is to ensure that all the required work and only the required work is clearly identified, that the deliverables and outcomes are documented, and that the boundary conditions are adequately defined. Scope planning involves referencing the project goals and objectives documented in the Project Charter or similar document. Scope planning is done in conjunction with schedule, resource, budget, and risk planning. During project implementation the tools in this chapter will provide a foundation for disciplined scope and change management and help to discourage scope creep.

SCOPE STATEMENT

The Scope Statement is a first step in planning a project. It helps inform subsequent tools in the planning and management effort. The Scope Statement is a written narrative of the project and product scope. It provides a written description of the project and each of the key deliverables. The project description elaborates on the high-level description in the Project Charter. Each of the key deliverables is described along with their criteria for acceptance. The Scope Statement also identifies the project boundaries, i.e., what is excluded from the project. If the project constraints are not itemized in the Assumption and Constraint Log (Chapter 3), they are identified in the Scope Statement.

This information defines the project and becomes the basis for making decisions and trade-offs during project planning and implementation. For organizations that define a scope baseline, the scope statement is a part of that baseline, along with the WBS and WBS Dictionary.

Project Management Toolbox: Tools and Techniques for the Practicing Project Manager, Third Edition. Cynthia Snyder Dionisio, and Russ J. Martinelli.
© 2025 John Wiley & Sons, Inc. Published 2025 by John Wiley & Sons, Inc.

Developing a Scope Statement

The quality of a Scope Statement hinges in many ways on the quality of the input information. Specifically, the following inputs carry great weight in developing a Scope Statement that has value:

- Project Success Criteria
- Project Requirements
- Project Charter.

Truly, a project's reason for existing is in helping an organization meet its business goals. Recognizing this, we reviewed earlier in the book several tools for capturing the business goals and project requirements Information from these tools are referenced when developing the Scope Statement.

When developing a Scope Statement it is important to include both project and product scope. Product scope encompasses the features and functions that will be delivered. Project scope encompasses the work to deliver those features and functions. All project management plan components and project documents are project scope. Products and components are product scope. For example, working software is an example of product scope. A test plan is an example of project scope. When you are delivering a service, the service is considered the product scope. Figure 6.1 shows a simple scope statement for a project.

Project Description

This section provides a narrative description of the work will you do to deliver the project outcome and support the business goals. It should be described in more detail than the charter, but without the technical detail that will be found in supporting requirements and specification documents. The scope statement provides context for the project deliverables that is not found in the WBS and requirements documentation.

Primary Deliverables

Performing the work described in the project description results in major deliverables or outcomes. Using the preceding example of the software project, the *major* deliverables are workflow analysis diagram, software designs, various plans and reports, and production software. A closer look at this set of deliverables reveals some guidelines for identifying the deliverables. First, there is a correspondence between elements of project description and the major deliverables. "Customer relationship workflow" (element of the project description) produces a "workflow analysis diagram" (deliverable).

Another guideline evident from the example is that the deliverables may include interim deliverables—for example, the prototype design, as well as end deliverables, the production software build.

Acceptance Criteria

Identifying deliverables is followed by defining the acceptance criteria, such as the expected elements or conditions for each deliverable. Some people choose to keep this information in a specifications document or a WBS Dictionary.

Scope Statement	
Project: Customer Relationship Software Implementation	
Project Description:	
Develop and deploy a customer relationship management software solution. Work to deliver the product includes an analysis of the customer relationship workflow, design the software solution, develop a prototype solution, perform customer usage tests, develop a final solution, conduct quality and acceptance tests, product documentation, and release of the software to the production platform. Project work includes team selection and leadership, stakeholder engagement, risk and issue management, and development and management of the schedule and budget.	
Primary Deliverables	**Acceptance Criteria**
Workflow analysis diagram	End-to-end workflow
Prototype SW design	Prototype software design that meets all identified requirements
Test plan	Test plan that includes unit testing, system testing, integration testing, and user acceptance testing
Customer user test report	Test report that summarizes the testing activities and results
Production software design	Production software design that incorporates all approved changes from the customer testing report
Quality test plan	Testing plan that identifies all metrics that will be tested and the acceptable result parameters
Quality test report	Test report that summarizes each test, the metrics, and the results.
Release plan	Plan that identifies how any outstanding items will be addressed, the implementation plan, and the follow-up support plan.
Production software build	Software build with all identified modifications and fixes implemented.
Documentation	A full set of categorized documentation that will be used for implementation, usage, and maintenance.
Constraints:	
Key developers will not be available in June due to travel to our European division. The budget is shall not exceed $200,000. Delivery date will be complete within 11 months.	
Project Work Exclusions:	
This project excludes any procurement work as this will be conducted by purchasing as needed. It also excludes work associated with customer- or customer-segment-specific customizations, field testing, and user training.	

Figure 6.1: Simple Scope Statement

Constraints

Constraints are factors that limit the project options. All projects face some constraints that influence the way project work is defined, deliverables are produced, and milestones met. These constraints may be physical, technical, resource, or any other limitation.

We discussed constraints as part of the Assumption and Constraint Log. Some people choose to only use an Assumption Log and keep their constraints in the Scope Statement. Because assumptions are updated frequently, they do better in a register as it is a dynamic document. However, constraints don't usually change and therefore can be documented in the Scope Statement.

Constraints are a foundational element of the project. It is likely you will need to rescope the project if the constraints change. Therefore, you should manage constraints by clearly identifying them and formulating the project around them.

Exclusions

It is an unfortunate truth that project managers consider anything not explicitly identified as in scope is out of scope. Conversely, stakeholders tend to consider anything not explicitly excluded as in scope. To avoid misunderstandings and conflicts you should specifically identify work that is explicitly excluded from the project scope. In the previous example Scope Statement, the project specifically excluded work associated with purchasing, customer requested customizations, field testing, and user training.

Using a Scope Statement

Except for highly repetitive projects that can utilize the same informal Scope Statement, every project regardless of the size and complexity can use a written Scope Statement. You will use it to develop and structure the WBS and make decisions based on what is in and out of scope. The acceptance criteria can be used to ensure alignment with requirements, specifications, verification, and validation. The exclusions will help to manage stakeholder expectations and reduce the likelihood of scope creep.

Variations

The Scope Statement that we have described is designed to be a cross-industry tool to serve as many project audiences as possible. Instead of using a Scope Statement, many government contract projects, or contractor/sub-contractor arrangements use a Statement of Work (SOW). A project Scope Statement and SOW serve related and often overlapping functions. Both set expectations and parameters for the project. The SOW, however, provides a detailed breakdown of the project and deals with the nuts and bolts details of how the project team will accomplish the project goals.

There are three basic types of SOWs:

1. Design/Detail SOW
   This SOW type tells a supplier or vendor *how* to do the work. The statement of work defines buyer or client requirements that control the processes of the supplier—such as measurement, tolerances, and quality requirements. In this type of SOW, the buyer or client bears the risk of performance.

2. Level of Effort or Time and Materials SOW

   This SOW type can be used for nearly any type of service. The real deliverable under this type of SOW is an hour of work and the material required to perform the work.

3. Performance-based SOW

   This SOW defines work agreements based upon what has to be accomplished. Desired outcomes from the work are described in clear, specific, and objective terms that can be reasonably measured.

A statement of work typically includes the following information about a project:

- *Project purpose.* Describes why the organization is investing in the project
- *Objectives.* Describes what the project aims to achieve once it is complete
- *Location of work.* Describes where the work will be performed
- *Period of performance.* Specifies the allowable time for projects. This can be represented in start and finish dates, number of hours that can be billed, or anything else that describes scheduling constraints
- *Deliverables.* Describes the tangible outcomes from the project
- *Deliverables schedule.* Describes when each of the deliverables is due
- *Value of work performed.* Describes the estimated total cost of the project
- *Specifications and standards.* Describes any specifications or industry standards that need to be adhered to in fulfilling the work obligation
- *Method of measurement of acceptance.* Describes how the buyer or receiver of the project outcomes will determine if the project outcomes are acceptable.

Whether a project manager chooses to use a Scope Statement, a Statement of Work, or some other variation, every project will benefit from the use of one of these tools. They provide the foundation for project planning as well as the overall vision of what the project end state will look like.

WORK BREAKDOWN STRUCTURE

A work breakdown structure (WBS) is a deliverable-oriented arrangement of the total scope of the project. It is arranged in levels hierarchically, with upper levels being decomposed into smaller, more detailed, and more manageable deliverables. The WBS establishes a visual framework for planning and managing the project work. It includes product and project work, providing the total project scope. Therefore, if work is not represented in the WBS, it is out of scope.

The WBS was initially used to bring order to the integration of management work faced with large and complex projects in the government domain. Today most projects in the business world are small and medium projects. Whether you are in software or hardware development, marketing or accounting, manufacturing, infrastructure technology, or construction, small and medium projects benefit from a WBS to structure the project work. While it is possible to run a successful project without using a WBS, experience suggests that probability of success is higher when you start with a solid WBS.

WBS Language

WBS level. The WBS is arranged in levels. Level 1 is the project. The first level of decomposition (Level 2) defines the organizing structure. Each lower level decomposes the work into more detail. There is no limit to the number of levels. Very large projects may have seven or more levels. Smaller projects usually have 3–4 levels.

WBS Component. Any entry in the WBS, that can be at any level.

Work package. A work package is a WBS component at the lowest level of the WBS. A work package can be used to develop estimates for cost and duration.

Control account. This is a summary work element that includes one or more work packages. It is used as a management control point where actual performance data is collected, analyzed, and reported.

Branch. All work elements underneath a level 1 deliverable constitute a branch. Branches may vary in depth.

WBS dictionary. A document that provides more detailed information about components in the WBS, such as activities, cost and duration estimates, and resources.

Scope Baseline. The approved version of the scope statement, the WBS, and the WBS dictionary.

Constructing a WBS: A Top-Down Approach

There are two basic ways of developing a project WBS: top-down and bottom-up. In this section we detail the top-down approach, which is a convenient approach for project managers and teams with experience in project work and knowledge of the project deliverables.

The development of a WBS is likely to be an easier and more meaningful exercise if you are equipped with information about the following:

- Project Scope Statement
- Project work
- Project requirements.

The Project Scope Statement provides an understanding of what the project will produce. You first need to know what you will produce (scope) before you decide how you will organize the work in the WBS. Note the practice of some experienced project teams who prefer to develop the Scope Statement and WBS in parallel rather than sequentially.

In constructing a WBS, the knowledge of the project work is crucial. For instance, to develop a meaningful WBS for a software development project, you need to understand the process of software development. Knowledge of the process will indicate the work necessary to produce the required project deliverables.

There are four aspects you need to decide on before constructing your WBS: the organizing structure, sizing guidelines, presentation, and a coding structure.

WBS Organizing Structure

Once you have the information about the scope, work, and requirements, the next step is to determine the organizing structure. There are three main methods you can use: project life cycle, key deliverable or system, geographic area (see Table 6.1).

The underlying principle of the project life cycle method is to use each phase of the project lifecycle as a branch in level 1 of the WBS. This approach of following the natural sequence of work is widely popular in some industries, such as software development. Common phases include requirements definition, high-level design, detailed design, coding, testing, and release.

Structuring a WBS by geographic sites or regions is sometimes used in construction or for rolling out a new product or service one location at a time. It is not unusual that literal geographic regions—for example, northwest site, southwest site, southeast site, and northeast site—are used as areas for level 1 of the WBS.

You have probably noticed that our discussion about the three structuring methods was limited to level 1 of the WBS. The lower levels can continue using the level 1 method, for example using geographic area for level 1 can have work elements on levels 2 and 3 that could comprise work elements that will be performed in the geography identified at level 1. Many practitioners find that a hybrid approach is more practical by combining two or three methods in the same WBS. For example, they may have a WBS by project life cycle on level 1, systems on level 3, and a geographic breakout of work at a lower level.

Sizing Guidelines

There are two aspects to sizing the WBS. The first is how many levels deep is your WBS, the second is what is the average size of a work package. The guideline for sizing is that you decompose until you have a work package that produces a tangible output that has enough detail to develop estimates and manage the work.

The project size is the key factor is establishing sizing decisions. A project that is approximately 6 months can probably get by with 4–5 levels. A project that costs over $10 million and is expected to take 3–4 years will likely have 6+ levels.

Table 6.1: Methods for Structuring the WBS			
	Method of WBS Structuring		
WBS Level	**Life Cycle**	**System**	**Geographic Area**
1	Project	Project	Project
2	Phase	System component	Area or region

Table 6.2: Examples of the Level of Detail of a WBS

(1) Project	(2) Project Duration (days)	(3) Project Budget (person-hours)	(4) # of Levels in WBS	(5) # of Work Packages	(6) Mean # of Hours/ Work Package (3) / (5)	(7) Mean # of Days/ Work Package[a] (2) / (5)	(8) Mean % of Budget/ Work Package [(6) / (3)] × 100
IT infrastructure	90	500	3	15	33	6	6.6
Selecting a billing platform	180	1200	4	36	33	5	2.7
S/W development	270	1200	3	25	48	11	4
Hardware development	365	500	4	29	17	13	3.4

[a]Assuming all work packages are sequential without any overlapping.

Work packages for small-to-medium-sized project may include work packages with 20–50 hours of work and 1–2 weeks in duration. Larger projects may include 80–200 hours of work, with each work package taking 2–4 weeks in duration.

In summary, establishing the level of detail of the WBS includes determining the number of levels in the WBS, the number of work packages, and the average size of the work package that are compatible with your tolerance level and the industry practice. Table 6.2 shows WBS data from some actual projects.

Coding Structure

Once a WBS gets beyond level 3, you will need a numbering or coding structure. You can use a numeric structure such as

1.0
1.1
1.2
1.n
2.0
2.1
2.2
2.n, etc.

You can also use an alpha-numeric structure by using the first letter of level 1 components followed by a number. For example, for a level 1 branch for Utilities, you would set up the coding structure as U, U1, U2, U3, U4, etc.

Presentation

Presentation refers to how you will display your WBS. Smaller projects that are up to 4 levels can use a tree structure, such as the one shown in Figure 6.2. After 4 levels, it starts to get confusing. Once you get to Level 5, it makes sense to switch to an outline presentation, such as shown in Figure 6.3.

Quality Check

Before you finalize your WBS, it is a good idea to give it a final review by checking it against these guidelines:

1. Make sure the WBS is deliverable-oriented. This means that the WBS components are nouns, not verbs. The potential exception is the Project Management branch which frequently represents level of effort type of work.
2. Check that the WBS includes all the work in the project, for both product and project scope.
3. Review the organizing structure to make sure it is logical.
4. Make sure that work packages are roughly the same size, unless you have branches of the WBS that will be decomposed at a later date.
5. Check that the lower-level components add up to the higher-level components with nothing missing, and no duplication.

Once you have completed your quality check, you are ready to share it with key stakeholders for sign off.

Increase Productivity Through WBS Templates

Having each project team develop a WBS from scratch can create several problems. First, WBS development consumes resources. Also, when each project uses a different WBS type, the benefit of comparing various projects and drawing synergies is lost. These problems have been successfully addressed by developing a WBS template for certain project families. Highway construction projects are an example of a project family. Other families may include software development projects, manufacturing process developments, and hardware development projects.

Generally, a family of projects is a group of projects that share identical or sufficiently similar project assignments. When the templates are developed and adopted, the development of a WBS for a new project starts with adapting the template. This saves time, produces a quality WBS, and enables interproject comparability. In a nutshell, templates increase productivity.

Tips for Using WBS Templates

Adopt a template WBS for each project family.
When developing the template, begin with few levels. Add more if project team members ask for it.
Build "blank" work elements in the template to be used for exceptions.
Start each project with a WBS template.
Know that smaller projects use a fewer number of levels in the template.

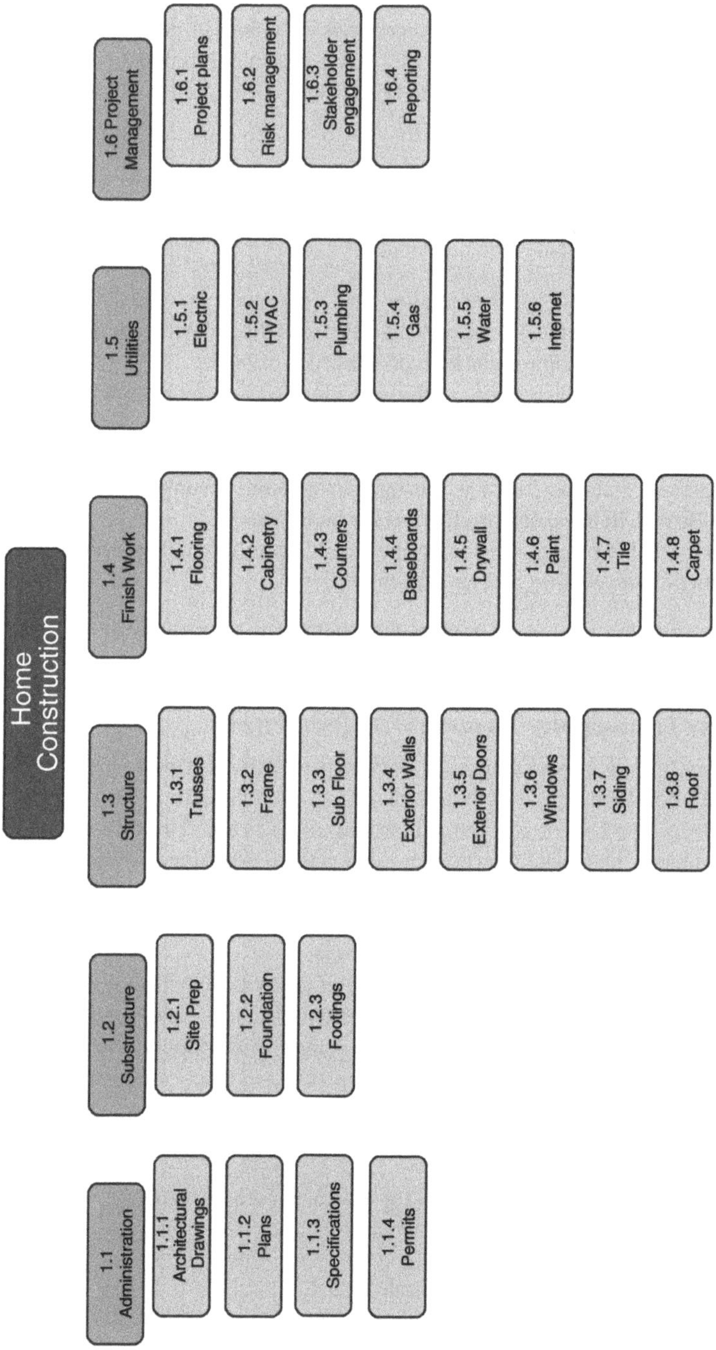

Figure 6.2: WBS in Tree Structure Format

Home Construction
1.1 Administration
    1.1.1 Architectural drawings
    1.1.2 Plans
    1.1.3 Specifications
    1.1.4 Permits

1.2 Substructure
    1.2.1 Site prep
    1.2.2 Foundation
    1.2.3 Footings

1.3 Structure
    1.3.1 Trusses
    1.3.2 Frame
    1.3.3 Sub-Floor
    1.3.4 Exterior walls
    1.3.5 Exterior doors
    1.3.6 Windows
    1.3.7 Siding
    1.3.8 Roof

1.4 Finish Work
    1.4.1 Flooring
    1.4.2 Cabinetry
    1.4.3 Counters
    1.4.4 Baseboards
    1.4.5 Drywall
    1.4.6 Paint
    1.4.7 Tile
    1.4.8 Carpet

1.5 Utilities
    1.5.1 Electric
    1.5.2 HVAC
    1.5.3 Plumbing
    1.5.4 Gas
    1.5.5 Water
    1.5.6 Internet

1.6 Project Management
    1.6.1 Project plans
    1.6.2 Risk management
    1.6.3 Stakeholder engagement
    1.6.4 Reporting

Figure 6.3: WBS in Outline Format

Constructing a WBS: A Bottom-up Approach

Brainstorming all project deliverables and then organizing them into a WBS hierarchy is at the heart of the bottom-up approach. This approach, essentially an application of the affinity diagramming method, is beneficial to those without much project experience. Projects developing or deploying novel technologies, typically fraught with high uncertainty and lacking precedents, can benefit from this approach as well, even if the project team is experienced. Despite its brainstorming nature, the bottom-up approach may be preceded by collecting necessary information for the WBS development, selecting the type of WBS, and establishing the level of its detail—in other words, some of the steps taken in the top-down approach. Other steps in the bottom-up approach follow.

1. *Generate a list of deliverables.* This step includes having each project team member brainstorm what is going to be created in the project. Each deliverable may be recorded on a sticky note and posted in a visible place.
2. *Sort deliverables into related groupings.* During this step you should also consolidate similar deliverables and eliminate redundancies. This should give you the preliminary WBS hierarchy.
3. *Name the groupings.* The name should reflect the type of deliverables that comprise the elements in the group.
4. Apply the quality checks identified for the top-down approach.

Using the WBS

The WBS is used as a framework for the project and as a foundation for all other areas in project management. The following list is a starting place for how the WBS is used in other project management areas:

Schedule. The WBS Structure is brought over to the schedule. The work packages are decomposed into the tasks necessary to create the deliverables.

Resources. The work packages are used to determine the types and amounts of physical resources required and the number and skill sets of the team resources required.

Cost. The work packages are used to develop detailed cost estimates for bottom-up budget development. Control accounts are used to track cost variances and cost performance.

Risk. Each work package can be assessed to determine threats and opportunities associated with the work package.

Procurement. Make or buy decisions are typically made at the work package or control account level.

Change. Changes to scope generally impact the WBS. The framework of the WBS can highlight how a change to one work package can impact other project or product scope.

PRODUCT BREAKDOWN STRUCTURE

For those project managers involved in product development, planning activities involve the combination of product-based planning and project-based planning. This type of planning is based upon the premise that the product solution is developed first, and *then* the project activities, tasks, deliverables, and resources required to create and deliver the product are planned.

The Product Breakdown Structure (PBS) is a critical tool for product-based planning and should be an essential part of any product development project manager's PM Toolbox. Essentially, the PBS disaggregates a final product into its constituent components.

Like the Work Breakdown Structure, the PBS is a visual aid that represents the relationship between a product and its sub-products. However, the project WBS and the PBS are used for different purposes. The main difference between the two is that the PBS focuses on the product, whereas the WBS focuses on the work required to create the product. That being said, if the WBS is arranged by key deliverable, the information from the PBS can be easily integrated into the WBS.

Constructing a Product Breakdown Structure

When creating a PBS, it is common to utilize the knowledge of the team which will be designing and developing the product. This usually involves a group of cross-functional product specialists. A structured brainstorming session, complete with a whiteboard, sticky notes, pens, and brainpower, is all you need to construct a PBS.

Start With the End in Mind

It is helpful to begin the PBS construction process by first diagramming the product which will eventually be created and developed. We recommend doing this by diagramming the *whole solution* that fulfills the customer's expectations.[1] For example, if we purchase a laptop computer, we wouldn't consider it acceptable if we received a box of circuit boards, a second box that contained the enclosure, another box containing the peripheral devices such as memory and network adapters, and finally an envelope containing a CD with the computer software applications and operating system. Rather, unless we're a computer hobbyist or a system integrator we *expect* to receive an integrated laptop that we can unpack, plug in, and begin using immediately. The point we're making is that the whole solution may include not only the core product, but also a number of enabling elements of the product that are needed in order to fulfill the customer's expectation.

It is helpful to think of the whole solution consisting of two parts: (1) the core components of a product and (2) the enabling components of a product. The core components are the tangible elements that when integrated, constitute the physical product developed. In systems language, these are the subsystems of the integrated product. The enabling components are the additional elements needed to ensure the product meets customers' expectations. Both the core and enabling components of a product need to be included in the planning of a project.

Let's look at an example to illustrate. Imagine that you are in charge of leading the project commissioned to create the next-generation smart phone for a leading phone manufacturer. Your whole solution diagram may look something like the one illustrated in Figure 6.4.

The product solution begins with the core components that consist of the physical elements that make up the phone such as the digital circuitry, the embedded software,

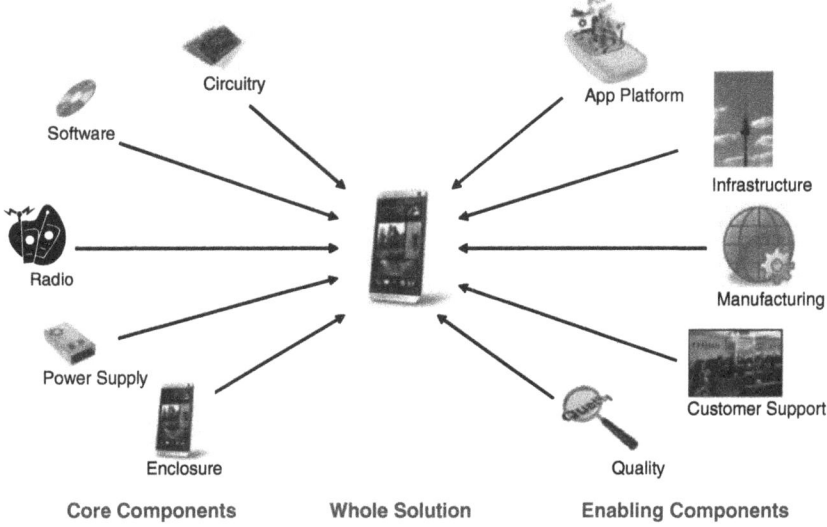

Figure 6.4: Example Whole Product Solution Diagram

the radio device, and the enclosure packaging (keep in mind this is a simplistic view for discussion sake).

The product solution also includes other important elements needed such as a software application development platform, interface to the wireless communication infrastructure, manufacturing of the product, quality assurance, and customer support for the users of the product. These are the enabling components of the product which are needed to ensure complete customer and user satisfaction when the phone is delivered.

By diagramming the product and its components you are beginning to structure your project. In effect, you are creating the project architecture, where the term *architecture* refers to the conceptual structure and organization of a system.[1] The process of creating a PBS now has an excellent starting point.

Choose the PBS Structure

Once you understand the core components of the product you intend to develop, you are ready to construct your PBS. The first step in this process is to select the type of structure you wish to use.

Like the project WBS, multiple structure types are available for the PBS—the hierarchical tree design, the outline design, and the mind-map design. The hierarchical tree and outline designs are identical to those described for the project WBS. If you create a diagram of the whole product solution as recommended in the previous section, the mind-map design is a good complementary PBS design to choose. Figure 6.5 illustrates an example of the mind-map PBS structure for the cell phone product discussed previously.

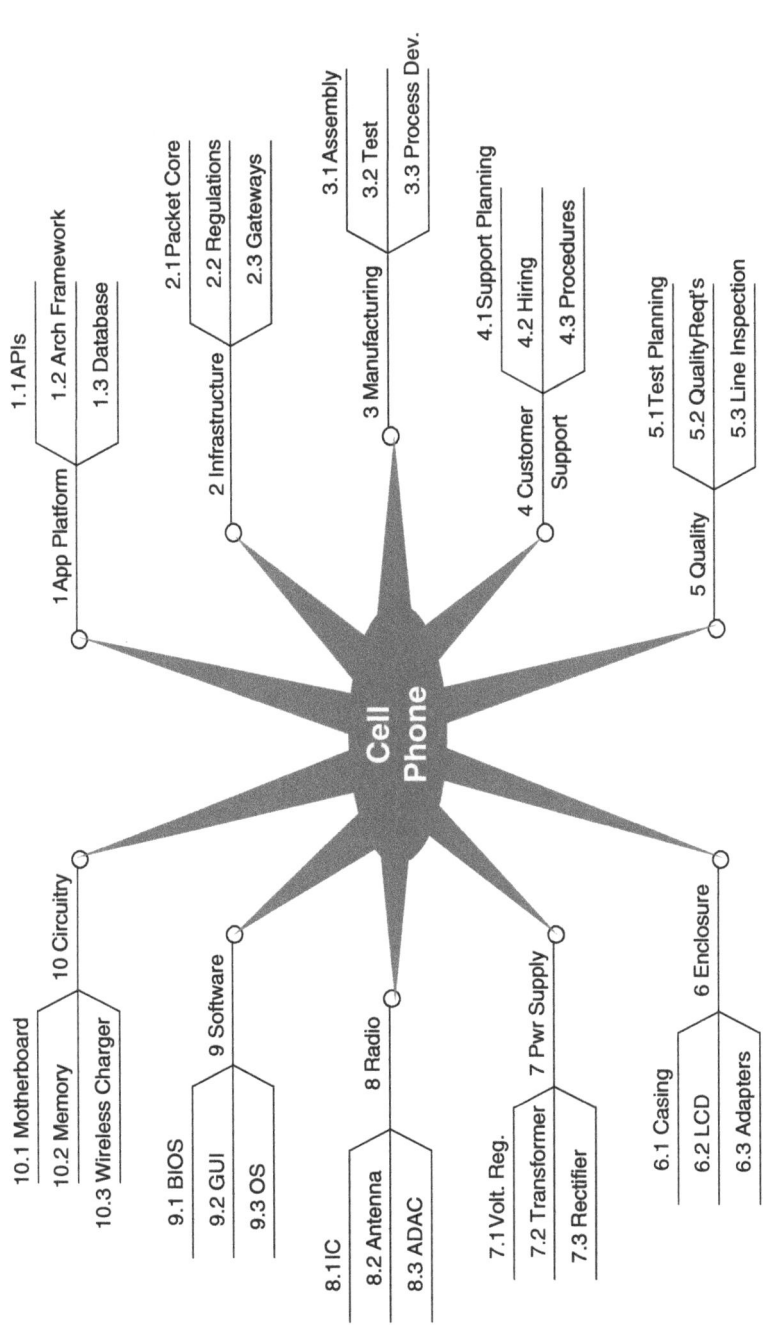

Figure 6.5: Mind-map PBS Design Example

Decompose the Product

With the PBS structure decided upon, the next step is to begin decomposing the product into its components and sub-components. At the top level (or in the middle of the mind-map structure) is the integrated product that will emerge from the product development process. In the example used earlier, the cell phone is the top-level product in the PBS.

Next, decompose the top level, or integrated product, into its primary components. As illustrated in Figure 6.4, you may have a combination of core components and enabling components if you take a whole solution approach. You will want to include both component types in your PBS. Continue decomposing each component into its subcomponents until a logical point of decomposition is reached (Figure 6.5). This is normally the level where the subcomponents can be described as a set of project deliverables.

Validate the PBS

Working your way up from the lowest level of the PBS, validate that the sub-components at each level will integrate to create the sub-component or component at the next highest level. Continue this process until the top-level product is validated. If gaps are discovered during the validation process, go back to the decomposition step to correct for the gap.

Using the PBS

As stated previously, the PBS is primarily used for product development projects where product planning and project planning are often occurring in a concurrent manner. The PBS is used to bring these two planning efforts together by providing a visual representation of the product and its components, as well as the relationship between the components, in order to facilitate the project planning process.

The PBS is used to describe *what* the project work effort involved is meant to create. This should be established before full-scale project planning begins, where project planning focuses on *how* the product will be developed.

Since planning on a product development project includes both product-based planning and project-based planning, once a PBS is created that decomposes a product into its components to the level of deliverables, the components can then become the elements contained within the WBS. Decomposition of the work necessary to create each product component can now take place. Used in this manner, the PBS and WBS become a set of integrated tools which demonstrates both the decomposition of the product and the decomposition of the work necessary to create the product.

WBS DICTIONARY

The WBS Dictionary supports the WBS by providing more detailed information about the work packages and/or control accounts that comprise the WBS. It is used in large projects to keep the work for each component of the WBS organized.

For Control Account Managers it functions as a compact collection of the information they need to create and manage the deliverables they are accountable for. While they need to have visibility into the overall project schedule and budget, their work is primarily contained in the WBS Dictionary.

When work is performed under contract, the technical information from the Statement of Work, the due dates, and the cost estimates (especially for a cost reimbursable contract), along with relevant contractual information, can be kept in the WBS Dictionary.

While the contents of the WBS Dictionary can be tailored to suit the needs of the project, they commonly provide a narrative description of the WBS element, associated tasks, resources, and cost estimates.

Creating a WBS Dictionary

The WBS Dictionary can be started once the WBS is complete. The first order of business is to determine the level of detail you want for the WBS Dictionary. As noted in the description of the WBS, a control account is a summary work element that includes one or more work packages. It is used as a management control point where actual performance data is collected, analyzed, and reported. A work package is a WBS component at the lowest level of the WBS. A work package can be used to develop estimates for cost and duration.

You may choose to record information at a control account level and keep information about the affiliated work packages at a high level. Conversely, you may want to store the details about work packages in the WBS Dictionary and indicate the control account they are associated with.

Determining the Contents

Once the level of detail is determined you can choose the contents you want to capture. We will assume a WBS at the work package level for purposes of this chapter. Typical contents include the following.[2]

Work package name. The work package name and number brought over from the WBS.
Description of work. A narrative description of the work package. This provides context and provides a deeper understanding of the work needed to complete the work package. Usually 3–5 sentences will suffice.
Milestones. List any key events associated with the work package. This can include start dates, approval dates, material delivery, and so forth.
Due dates. The due dates associated with the milestones.

Detailed information about activities can be kept in a table format, as shown in Table 6.3 to keep the information well organized and easy to absorb.

ID. Record an Activity ID by extending WBS numbering scheme. For example, a work package with a number of 1.4.2 would have activities numbered 1.4.2.1, 1.4.2.2, and so forth.

Table 6.3: Activity Information for a WBS Dictionary									
			Labor			Material			
ID	Activity	Resource	Hours	Rate	Total Labor	Units	Cost	Total Material	Total Cost

Activity. List the activities associated with the work package. These are generally obtained from the project schedule.

Resources. For each activity, list the resources needed to complete the work. This includes both team resources (labor) and physical resources (material).

Labor hours. The estimated effort hours for each resource. Generally effort is recorded in hours of days. Effort indicates the amount of work it will take, not the duration of the work.

Labor rates. The labor rate, per hour, day, or flat rate. This information can be acquired from the cost estimates.

Total labor. The sum of the labor hours times the labor rates.

Material units. The estimated quantity of each type of material used to complete the activity work. Material includes supplies, equipment, licenses, location costs, and so forth.

Material costs. The cost per unit, whether it is per square foot, per gallon, flat rate, or some other measure.

Total material. The sum of the material units times the material costs.

Total cost. The sum of the total labor costs and the total material costs.

Some project managers find it useful to include additional relevant information associated with a work package either in the WBS Dictionary directly, or by referencing or linking to other documents. Examples of additional information include quality requirements, contractual information, technical information, and acceptance criteria.

Collecting the Information

The WBS Dictionary is generally started when the WBS has been signed off on. However, the information needed to populate the WBS Dictionary takes time to develop. The WBS Dictionary uses information from the schedule, resource requirements, and cost estimates. Depending on the contents, it may also require technical specifications, contract statements of work, and other information. Therefore, it is a dynamic document that is updated as information becomes available.

Most projects use some form of progressive elaboration, where more detailed information is developed as needed. As that information is recorded in the schedule, budget, or other documentation, that information is entered into the WBS Dictionary. Scope changes can necessitate an update to the WBS dictionary as well.

Using the WBS Dictionary

The project manager uses the WBS Dictionary to organize information by work package. Large projects may have Control Account Managers (CAMs) who are accountable for all the work associated with one or more control accounts. When this is the case, the CAM develops, maintains, and uses the WBS Dictionary. In essence the WBS Dictionary functions as a mini-project plan for the control account. The scope, resource, schedule, and cost planning are all recorded in this document. Status from each control account is collected and reported up to the project level. Therefore, the information in the WBS Dictionary is vital to planning and tracking project performance.

References

1. Martinelli, R., Waddell, J., and Rahschulte, T. (2014). *Program Management for Improved Business Results*, 2nd ed. Hoboken, N.J.: John Wiley and Sons.
2. Dionisio, C.S. (2023). *A Project Manager's Book of Templates*. Hoboken, NJ: John Wiley and Sons.

7

SCHEDULE DEVELOPMENT

Project scheduling involves the planning of timelines for completing the work and establishing dates when project resources will be needed to perform the work.

The project schedule is the cornerstone of project work, and as such, serves as the working tool for project planning, performing, and tracking. By developing a project schedule, a project manager is *planning* the time element of the project. Comparing the actual dates of the tasks with the scheduled dates, a project manager *tracks* the performance and when actual performance deviates from the scheduled dates the project manager can take corrective action.

The process of schedule development involves the integration of multiple aspects, including the ensuring the needed resources are available, estimating the effort and duration of tasks, and balancing the constraints imposed by the budget and expected due dates. The goal of schedule development is to be able to answer the following questions:

1. If performed to plan, when will the project be completed?
2. What is the best way to sequence the work?
3. When should each task begin and end?
4. Which tasks are more critical to ensure timely completion of the project?
5. Which tasks can be delayed, if necessary, without delaying the project completion date?
6. When are project resources needed and when will they be released?

The project schedule can be presented in various ways as demonstrated in this chapter. The schedule type is often driven by the preferences and needs of the various project stakeholders. A functional manager, for example, may be interested in a schedule that shows the resource allocation requirements, the senior sponsor may only be interested in a schedule that shows major project events and milestones, and a project manager may need a detailed schedule that shows the start and finish of each task.

Since project managers have a set of stakeholders with varying scheduling needs, they need to have various schedule types in their PM Toolbox. We will start by looking at a high-level schedule, the milestone chart.

Project Management Toolbox: Tools and Techniques for the Practicing Project Manager,
Third Edition. Cynthia Snyder Dionisio, and Russ J. Martinelli.
© 2025 John Wiley & Sons, Inc. Published 2025 by John Wiley & Sons, Inc.

MILESTONE CHART

A Milestone Chart is a visual scheduling tool that shows milestones against a timescale to signify key project events. A *milestone* is defined as a significant event or point in time. For instance, *Requirements Document Complete* is a distinctive milestone for software development projects, and *Market Requirements Document Complete* is a characteristic milestone for product development projects. While these milestones relate to the completion of key deliverables, other types may include the start and finish of major project phases, major reviews, events external to the project (e.g., trade show dates), and so forth.

As shown in Figure 7.1, the Milestone Chart effectively communicates the planned dates for significant events and can also be used for tracking to indicate of there are any slippages.

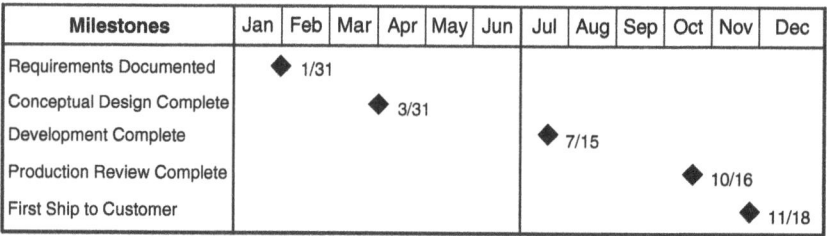

Milestones	Jan	Feb	Mar	Apr	May	Jun	Jul	Aug	Sep	Oct	Nov	Dec
Requirements Documented	◆ 1/31											
Conceptual Design Complete			◆ 3/31									
Development Complete							◆ 7/15					
Production Review Complete										◆ 10/16		
First Ship to Customer											◆ 11/18	

Figure 7.1: Example Milestone Chart

Who Is Involved in Scheduling?

The involvement of project participants in developing schedules hinges to a great extent on the organizational strategies for project management. In the matrix environment, for example, team members, project managers, functional managers, and the project office may be involved in developing the schedule.

Team members typically own work packages and tasks, reporting their completion and estimating how much time is necessary to complete the work. While they have to know some scheduling terms, such as start date, finish date, data (reporting) date, and resource availability, there is no need for extensive knowledge of scheduling theory.

As providers of resources to a project, functional managers care about the accuracy of the estimates and availability of resources when projects need them. Like team members, their knowledge of scheduling theory is basic.

Project managers are the ultimate users and owners of the project schedule. They facilitate schedule development and monitor the data furnished by team members for completeness and feasibility. They then use the schedule to manage and track progress to plan, working with the functional managers to make schedule modifications when needed. Project managers need a decent amount of knowledge about scheduling theory.

The project office (or the scheduling group) should have scheduling experts who are capable of designing and maintaining a project scheduling system that all other players utilize. Also, their knowledge in running scheduling software and checking time, cost, and resource estimates in order to support the system and individual projects is essential.

Developing a Milestone Chart

Traditionally, the Milestone Chart has been used to focus senior managers on important project events, whether the project is large or small. It can also help strengthen the emphasis on goal orientation while reducing focus on task orientation.

The first encounter with milestones usually occurs with the Project Charter, or other similar start-up document. Milestones in these documents may not be very accurate and are subject to updates as more information about the project is known.

As the project manager and the project team gain a solid definition of the project scope, the quality of the information on the Milestone Chart improves. When a detailed schedule has been developed, the Milestone Chart may again be updated, and at this point it is considered quite accurate.

Depending on the needs of the project, you may keep the milestones at a high level, such as shown in Figure 7.1. This type of milestone chart is intended to inform managers or other stakeholders about planned dates for key events. High-level charts may only identify start and finish of the project and its major phases.

Another option is a Milestone Chart designed to help manage the work necessary to accomplish a project deliverable or complete a project phase. This type of chart is used by the project team to focus them on major project events or deliverables. Information in this type of Milestone Chart would include the same information as the high-level chart, as well as other types of milestones, such as key deliverables, major reviews, important events external to the project, and so forth. This type of milestone chart helps focus on the major synchronization points and key decision points in the project.

Once you have identified the key milestones and along with their dates, list milestones on the vertical axis of a sheet, show the timescale across the horizontal axis, select symbols for milestones (diamond, for example), and place the symbols across the timescale (see "Tips for Milestone Charts").

Tips for Milestone Charts

- Because milestones mark a significant event, they have a duration of "zero." Performing work to reach a milestone has a duration, but the actual accomplishment of the work is depicted as zero.
- When building a milestone chart in scheduling software, you can enter 0 days for duration and the software will automatically turn it into a milestone.
- When the team is involved in establishing the milestones and the schedule, there is broader buy-in and stronger commitment from team members.

You may choose to show connections between milestones, or not. Once you have the first draft, ask yourself: Are all necessary milestones there? Are they logically sequenced? In appropriate positions on the schedule? It is a good practice to ensure there is no large gap in between milestones, since having large gaps reduces the ability to manage progress to set milestones.

Using the Milestone Chart

Milestone charts are used to communicate high-level information to key stakeholders. Stakeholders like senior management want to know if the project is on schedule or falling behind. They are not interested in the details found in a resource-loaded schedule, or whether a task begins or ends on time. They want a quick way to determine if the project work is performing according to plan.

The Milestone chart with few entries and specific dates communicates whether a milestone was met or missed. For missed milestones you can drill down into a more detailed schedule to determine the cause of the slippage and take corrective action.

NETWORK DIAGRAM

A network diagram is a visual display that shows the relationships between project work. It is sometimes referred to as a precedence diagram. The network diagram is used to plan and analyze the flow of work through the project. It can be shown at various levels, such as at the work package or the detailed task level. The network diagram generally indicates work as nodes that are connected by arrows. The beginning and end of the network diagram is shown as a milestone, as indicated in Figure 7.2.

Understanding the nature of work (hard logic or preferential logic) and the sequence of events provides the structure that the rest of the schedule will be built on.

Developing a Network Diagram

The network diagram is developed by identifying dependencies between tasks. This means determining a task's immediate predecessor and immediate successor. Some

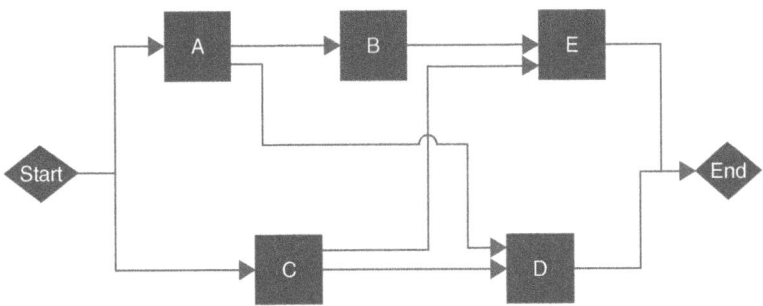

Figure 7.2: Example Network Diagram

dependencies are based on hard logic, some by soft logic, and others by resource constraints.

Types of Dependencies

Hard logic means that the dependencies are determined by the nature of the work. An example is that you must write the code before testing it; the other way around is not possible. Soft dependencies (also known as preferential dependencies) are not required by the work logic but reflect a best practice or one's experience and preferences. For example, we may decide to write a piece of software code, test it, write another piece, test it, and so forth. Conversely, you could do all the coding before starting the testing. This decision is based on preferential or soft logic. Dependencies may also be dictated by availability of key resources. If two tasks require the same resource, one will have to follow the other. Once the dependencies are established, they can be recorded, as shown in Table 7.1.

Dependency Relationships

Most relationships between tasks show that one task must be finished before the next one can begin. This is known as a finish-to-start relationship. All the relationships in Figure 7.2 are finish-to-start. This type of relationship is abbreviated as FS for finish-to-start. However, there are other types of relationships, specifically start-to-start (SS), finish-to-finish (FF), and start-to-finish (SF).

Table 7.1: Network Diagram Information	
Task ID	**Predecessor**
A	n/a
B	A
C	n/a
D	A, C
E	B, C

Figure 7.3: Start-to-Start
with a Lag

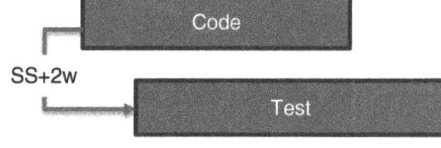

Figure 7.4: Finish-to-Finish
with a Lag

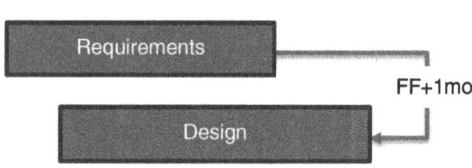

Figure 7.5: Finish-to-Start
with a Lead

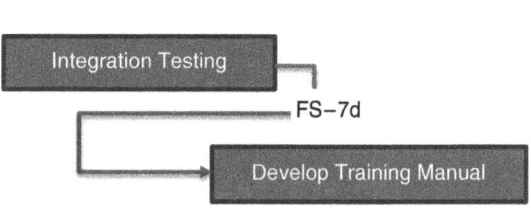

In a start-to-start relationship the predecessor task must start before the next task can start. The first task does not have to finish before the next task begins, it just has to start. For example, if you have one person coding the software, and another person testing it, the tester can start testing sometime after the coder starts coding. This type of relationship often includes a lag. A lag is a directed delay between tasks. Thus, you might see a relationship between coding and testing as SS + 2w where 2w means 2 weeks. This means that two weeks after the coder starts their work, the tester will begin. Figure 7.3 shows this relationship.

The finish-to-finish relationship indicates that the predecessor task must finish before the successor task can finish. You will want to finish gathering requirements before finishing with the design work. You do not need to finish gathering all the requirements before starting work on the design, but you can't finish the design until all the requirements are known. Figure 7.4 shows this relationship with a one-month lag.

There are times when you want to accelerate the relationship between tasks. In these situations you would use a lead to indicate an acceleration. A lead is most often used with a finish-to-start relationship to indicate that the successor task can start sometime before the predecessor task finishes. For example, you may start creating a training manual for the new software before the integration testing is complete. Figure 7.5 shows a finish-to-start with a 7-day lead.

Now that you have learned about each of these relationships, Figure 7.6 shows what they look like on a single network diagram.

Knowledge of the different relationships between tasks, as well as the ability to accelerate work with a lead or delay work with a lag allows you to fine tune the schedule.

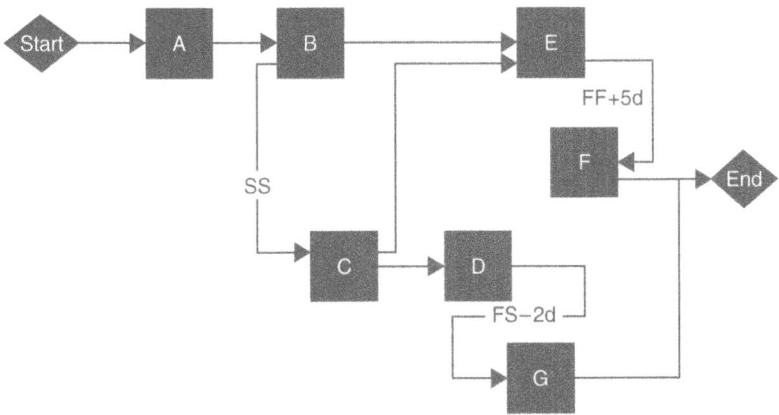

Figure 7.6: Network Diagram with Multiple Relationships

Tips for Network Diagrams

You must understand the nature of the work you are doing to build an accurate network diagram. You can't just assume you can overlap tasks with leads or accelerate work by making a finish-to-start relationship a start-to-start relationship.

It is a good practice for 90% of your relationships to be finish-to-start. That is usually the nature of how work flows on a project. When you start to manipulate the flow of work too much with different types of relationships, or an excessive use of leads and lags, you increase the risk of rework.

Network diagrams do not account for resource availability. They only show how work can flow. Ultimately the flow of work depends on a lot of variables, including resource availability.

Using a Network Diagram

There are several uses for the network diagram. The first is to analyze the sequencing of work. This is usually done for small chunks of work to identify the most efficient way to get things done. For example, if you are replacing all the computers, printers, and other peripherals for 500 users, you might work out the flow first, before committing it to software. For chunks of work like this you can use post-it notes on a white board to move tasks around until you find the best option.

Another use lies in sequencing the tasks in scheduling software. This provides the foundation of the schedule. You will need effort and duration estimates along with resource availability and reserve before your schedule is complete, but the network diagram is the first step in creating your schedule.

There are times you need to compress the schedule, either to meet a predetermined milestone or to make up for a schedule slip. You can employ network diagramming to see where you can overlap tasks or change the dependency type to reduce the time it takes to get all the work done. This is called fast tracking. While fast tracking is useful, it can add risk to the schedule, as most work is best done sequentially. However, used in small doses, fast tracking can reduce the time it takes to complete a phase or a project.

CRITICAL PATH METHOD

The Critical Path Method (CPM) tool builds on the Network Diagram technique for analyzing, planning, and scheduling projects. It provides a means of determining the longest series of tasks through the project. This path is called the critical path. Any delay on the critical path equates to a delay on the project completion date.

When planning a project, a CPM Diagram helps the project manager to see the total completion time, understand the sequencing of tasks, and ensure resources are available when necessary. When executing the project it helps the project manager monitor those tasks that are critical, measure progress, and identify tasks that can be fast tracked or crashed if you need to make up time.

Constructing a CPM Schedule

Constructing a CPM schedule involves several steps.

1. *Identify tasks.* Generally, team members do this by decomposing the deliverables from the WBS.
2. *Sequence the tasks.* The output of sequencing the tasks is the network diagram described in the preceding topic.
3. *Allocate resources.* Resources include internal team resources, contract resources, and physical resources. For team and contractor resources you will need to know the skill level and availability of resources. Using a Responsibility Matrix (described in the chapter on Resource Planning) is helpful when allocating resources.
4. *Estimate task duration.* The duration of each task is dependent on the nature of the work and the availability of resources. Chapter 9 describes several methods to estimate. The information in that chapter describes cost estimating, but the same techniques can be used for effort and duration estimating.
5. *Analyze the schedule.* This is where you first see the critical path and the overall duration of the project. At this point you may choose to work with the task sequencing (as described in the Network Diagram topic) to compress the schedule. You should also consider where and how much schedule reserve to include (reserve is addressed in the chapter on Risk).
6. *Baseline the schedule.* Once you have resource commitments, schedule reserve, and sign-off, you baseline the schedule. This is the schedule you will use to measure progress throughout the rest of the project.

To avoid repetition, we will assume that the work from the first four tasks has been done.

Why a Team Approach to CPM Development?

Using a project team to help with the CPM diagram is perhaps the most effective way of doing it. Here is why:

Team members are usually the best source of knowledge about their piece of the schedule.
Each team member can see where and why he or she is critical to the success of the project.
The team can find creative ways to best sequence and shorten the duration of tasks and the total project.
As a unit, the team can focus its energy and mind on mission-critical tasks.
Involvement of team members enhances commitment and a sense of ownership of the project.

Table 7.2 shows the network diagram information from Figure 7.6 with durations added.

Figure 7.7 shows the resulting network diagram with the durations added. Even though this network diagram only has 7 tasks, it is challenging to determine which path through the network will take the longest, i.e., the critical path. Imagine how daunting it would be to figure that out if you had 100 or more tasks. Fortunately, scheduling software takes care of that. However, it is good to know what is happening in the software to calculate the critical path. That way, if you look at a schedule and think something is off, you know how to identify issues, such as an illogical network diagram, a missing dependency, or other issue.

To understand what the scheduling software is doing, we have to introduce some new terms.

Table 7.2: Network Diagram Information

Task ID	Predecessor	Duration
A	n/a	5
B	A	12
C	B Start-to-Start	7
D	C	8
E	B, C	15
F	E Finish-to-Finish + 5d	10
G	D Finish-to-Start – 2d	9
End	G, F	

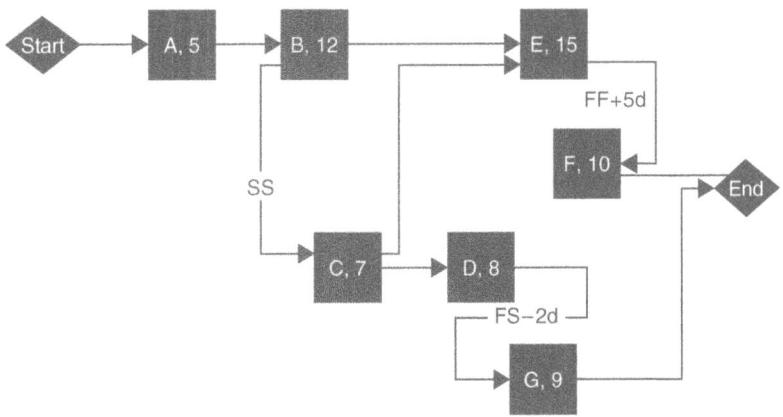

Figure 7.7: Network Diagram with Durations

Critical Path Terminology

CPM has several terms you need to know.

Early start. The earliest a task can start based on the network logic.

Early finish. The earliest a task can finish based in the early start and the duration.

Late start. The latest a task can start based on the late finish and the duration.

Late finish. The latest a task can finish without causing the project to be late, based on the network logic.

Total float. The amount of time a task can slip without causing a delay in the project finish date.

Free float. The amount of time a task can slip without causing a delay in the next task.

Identify the Critical Path

The diagram shows a number of different paths from Start to Finish. To calculate the time to pass through a path, add up the times for all tasks in the path. The critical path is the longest path (in time) from Start to Finish. It indicates the minimum time necessary to complete the entire project.

ES–EF

While adding up task times is simpler for smaller projects, it is too cumbersome and difficult for larger projects. For larger projects you will need to conduct a forward and backward pass to determine the critical path. Say, for example, you have the start date for a project. Then, for each task there exists an earliest start time (ES). Assuming that the duration of the task is D, then its earliest finish time (EF) is ES + D. Figure 7.8 shows a template for how each task in the network is shown for a manual critical path calculation.

Figure 7.8: Task Template for Critical Path
Calculations

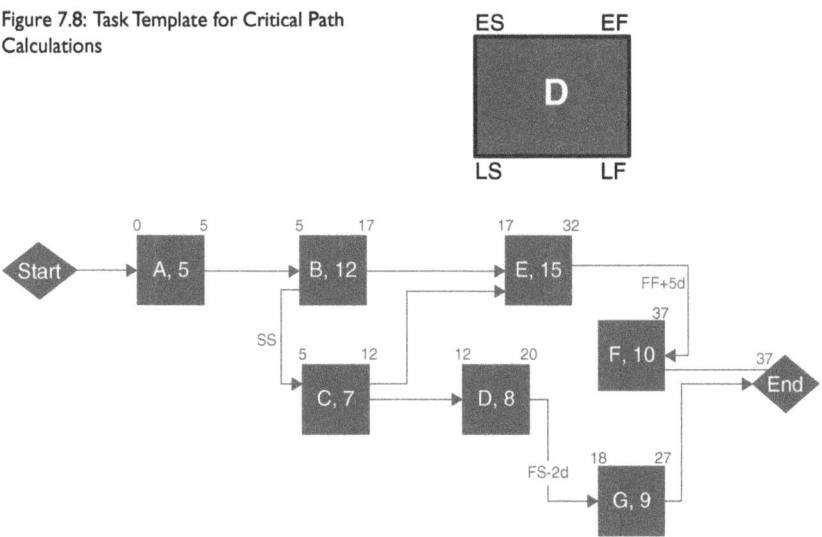

Figure 7.9: Network Diagram with Forward Pass

The first step in identifying the critical path is to go through the network from left to right, starting with the first task. This is called a forward pass and is used to calculate the ES and EF for each task. The process is as follows:

- ES for the first task is 0
- EF for the first task is $0 + D = EF$
- ES for following tasks is the largest (or latest) EF of any immediate predecessor tasks.

Thus, if you have a duration of 5 for the first task and 3 for the second task, you have the following:

- ES for Task 1 is 0
- EF for Task 1 is 5
- ES for Task 2 is 5
- EF for Task 2 is 8

Figure 7.9 shows the network diagram with durations and the forward pass completed.

The backward pass is calculated going through the network from right to left, starting with the task with the highest early finish. If you want to finish the project by the date of the EF for the project, the late finish (LF) of the last task is the same as the EF of the last task. The LF for any task is the latest time a task can finish without delaying the project.

You calculate the late start (LS) as $LF - D$. Building on these concepts, we can go through the backward pass to calculate the LF and LS for each task (see Figure 7.10):

- LF of the last task is either the EF or a predefined fixed end date
- LS for the last task is $LF - D$
- LF for previous tasks is the smallest (or earliest) LS of any of immediate successors.

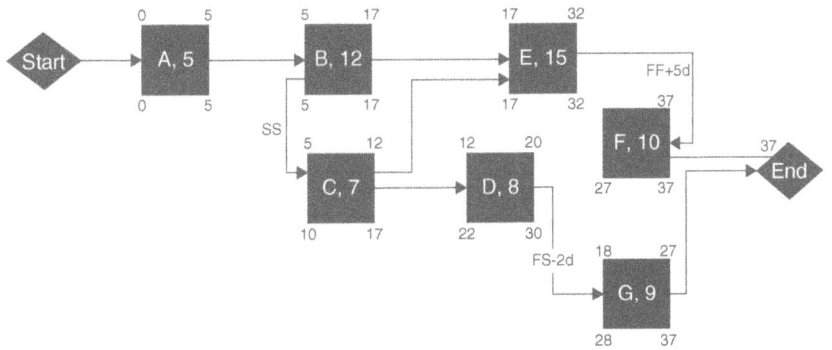

Figure 7.10: Network Diagram with Backward Pass

Now that the forward and backward passes are finished, note that Figure 7.10 shows that in some tasks, early start is equal to late start, while in some it is not. The difference between a task's early and late start (or early and late finish) is called *total float* or just *float*. Total float is the maximum amount of time you can delay a task beyond its early start or finish without delaying the project completion time. Figure 7.10 indicates that C, D, and G have total float. C has 5 days of total float, and D and G have 10 days of total float.

Free float is another kind of float equal to the amount of time you can delay a task without delaying the early start of any immediately following task. A task with the positive total float may or may not have free float; however, the latter never exceeds the former. The formula to calculate free float is the difference between the task's early finish (EF) and the earliest of the early start times (ES) of all of its immediate successors.

In our example in Figure 7.11, only task G has free float, while all other tasks have zero free float. This means that the only task that can start later than the early start, without impacting the ES of the next task, is G.

Tasks on the critical path have zero total float and are called *critical tasks*. They are shown in Figure 7.11, with thick arrows indicating tasks on the critical path. Occasionally you will have multiple critical paths; this makes managing timely delivery more challenging.

A task with zero total float has a fixed scheduled start time, meaning that ES = LS. Consequently, to delay the start time is to delay the whole project, which is why such tasks are called *critical*. In contrast, tasks with positive total float offer some flexibility. For example, you can relieve peak resource usage in a project by shifting tasks on the peak days to somewhere between the early and late start dates. That won't impact project completion time. In case of the free float, you can delay the task start by an amount equal less than the free float without affecting the start times of succeeding tasks.

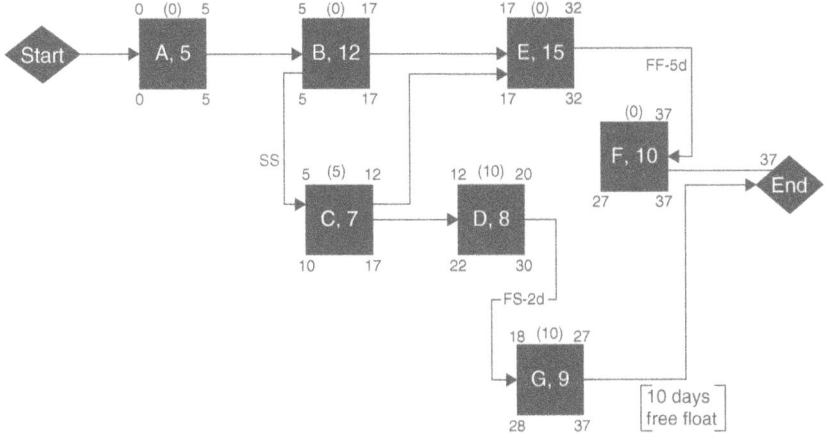

Figure 7.11: Critical Path and Float

Analyzing the Network

Once you have completed your forward and backward passes and identified the float, look closely at the diagram and ask the following questions:

- Has any important task been left out of the schedule?
- Is the task sequencing logical?
- Are durations of tasks reasonable?

If a project needs to be completed as soon as possible with less regard for resource availability, the project manager must check if it is possible to reduce the project duration. The only avenue to do that is to find ways to shorten tasks along the critical path. This is possible by fast tracking, crashing, or a combination of the two approaches. Note that fast-tracking or crashing noncritical tasks is irrelevant, because it does not reduce the duration of the critical path.

Fast tracking means changing the hard and soft dependencies—in other words, changing the logic of the diagram by attempting to overlap tasks on the critical path. In the process, neither task durations nor resource allocations will be changed.

Crashing means shortening the duration of tasks along the critical path without changing dependencies. You do this is by assigning more people to the tasks, working overtime, using different equipment, and so on. The crucial question is whether the gains from the reduction of project duration exceed the costs of acceleration. For the majority of time-to-market projects the answer to the question is yes.

There is more information on schedule compression in Chapter 14.

Using the CPM Diagram

The CPM tool was originally developed for large, complex, and cross-functional projects because it can easily deal with a large number of tasks and their dependencies. This is the method that most scheduling software uses.

Scheduling software can show the CPM information in a network diagram with dates or a Gantt Chart. Once you load resources into the scheduling software you will see a change in many of the dates. This can happen for a number of reasons:

- If you have multiple resources on one task, the duration will be reduced.
- If you have resources that only work part time on a task the duration will be increased.
- If you have resources that are assigned to multiple tasks on the same dates, you will need to resolve the over-allocation by shifting the start and end dates within the available float.
- If you can resolve over-allocation by shifting the start and end dates, you will need to extend the date of the project or find more resources with the proper skill sets.

When created in a robust scheduling software, developing a hierarchical schedule using rolling wave planning produces a schedule that can be used at multiple levels. A medium-sized project might show two levels, a milestone chart and the detailed tasks. A large project will be better off with three levels, milestone, summary tasks, and detailed tasks.

Rolling Wave Planning

When starting some projects, we only have information about an early phase, while details about later phases emerge as the project progresses. In response, at the start of the project, we can develop a high-level schedule encompassing the whole project and then build detailed schedules of the project's major phases as details become available. This is known as rolling wave planning.

With rolling wave planning the project team is not forced to develop a schedule for tasks for which they have little or no information at the current moment. Rather, they can build a flexible big picture schedule, focusing on near-term tasks first, and then add longer-term detail as the project progresses through the project cycle.

Level 1 is a summary schedule of the project, which is usually a milestone chart format, as shown in Figure 7.1. It is an outline that will be used throughout the project as a tool to report progress to top management. It is important to highlight events that need critical attention, such as material requirements, vital tests, and completion dates.

The Level 2 schedule will progressively elaborate tasks from the master project schedule, showing them in more detail. This schedule is generally used to assign responsibilities for work packages. It enables you to scrutinize and develop the high-level network diagram that establishes the structure of the project and identifies dependencies between various phases and milestones.

Level 3 is a detailed schedule intended to show work at the weekly or daily level. It is most often shown as a Gantt chart, or in Hybrid project, you may see it shown as a Task Board developed from the backlog.

CRITICAL CHAIN SCHEDULE

Introduced in 1997, the Critical Chain Schedule (CCS) is a network diagram that strives for accomplishment of drastically faster schedules (see Figure 7.12). It uses several unique approaches.

1. The CCS focuses on the critical chain, the longest path of dependent events that prevents the project from completing in a shorter time.
2. Task durations are estimated with 50% probability. For this reason, they are significantly shorter than those used in other scheduling tools, which are often with 80–90% probability.
3. In contrast to the critical path, the critical chain is defined by the resource dependencies.
4. Buffers, such as feeder buffers, resource buffers, and project buffers, are built to protect the critical chain in the course of project implementation.

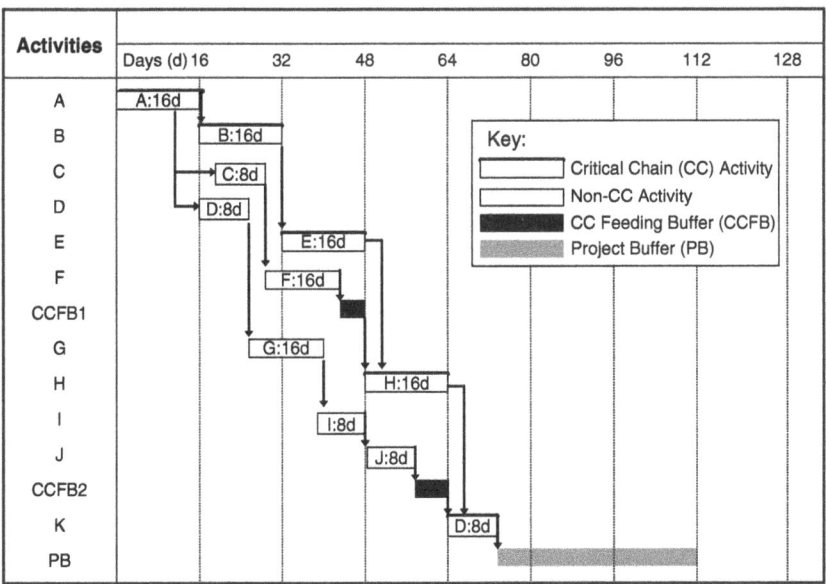

Figure 7.12: Example Critical Chain Schedule

Developing a Critical Chain Schedule

One of the differentiating factors in a CCS is the emphasis on dedicated team resources, meaning that team members work full-time on one project only. Because of this, members of the dedicated project teams are more productive than members who are shared by multiple project teams. A reason for this is the switching time cost created by one's work in multiple projects is eliminated. Although this is generally true, there are some exceptions. A study found that when a team member who is focused on a single project is assigned a second one, productivity often increases a bit because the team member no longer has to wait for the tasks of other members working on the initial project (see Figure 7.13). Rather, the team member can float back and forth between the two projects.[1]

When a third, fourth, and fifth project is added, however, the productivity plummets rapidly and the team member becomes a bottleneck of all projects he or she is involved in. This is why the CCS approach insists on using dedicated teams.

Developing a CCS starts with identifying and sequencing tasks, similar to the CPM method. Where CCS and CPM differ is in the area of resources and durations.

Resources and Task Durations

As mentioned previously, in critical chain projects, resources are allocated 100% of the time. The key difference is that when estimating durations CCS does not allow for contingencies. When estimating how long a task will take, most people allow for things that could go wrong, rework, and delays. In other words, they add some contingency to their individual duration estimates because they don't want to deliver late. Critical chain estimating makes a case for truth in duration estimating[2] by asking people to estimate what they think the actual hands-on time (AKA touch time) is for a task. The duration estimates are expected to have a 50/50 chance of being sufficient for the work. This results in a much shorter estimated project duration.

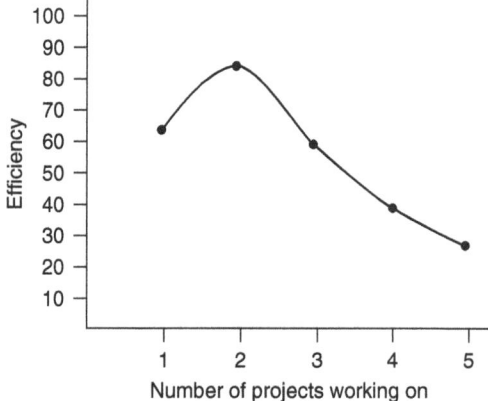

Figure 7.13: Productivity of Multi-project Team Members

Identify the Critical Chain

The critical chain is the longest path in the network diagram, considering task and resource dependencies. Or stated a different way, it is the sequence of dependent events that keeps the project from completing in a shorter time.

Add the Resource Buffers

Critical Chain Schedules always consider the resource constraints and include the resource dependencies that define the overall longest path. This is addressed by adding resource buffers to protect the critical chain from unavailability of resources. Resource buffers are added to the critical chain only, do not take any time in the critical chain, and are termed *resource flags*.

For example, any time a new resource will be used in a critical chain task, you add a resource buffer. This signals to the project manager and resource provider when to make the resource available to work on a critical chain task.

Create a Project Buffer

Unlike other schedule development tools, the CCS uses a project buffer. Its purpose is to protect the project completion date by aggregating risk contingency time in the form of the project buffer at the end of the critical chain. There are several methods to determine the buffer duration. One of them is to divide the duration of the critical chain by two (called the "50% buffer sizing rule"). The buffer is used to absorb uncertainty or disruptions that may occur on the critical chain and has no work assigned to it (see Figure 7.12).

Create Feeding Buffers

Protecting the critical chain with the project buffer is not enough. There is a significant risk that tasks that are not on the critical chain but feed into it may slip to the point of pushing out the critical chain. To protect the critical chain from the risk, we can aggregate contingency time at all points where non-critical tasks feed into the critical chain (see Figure 7.12). These contingency times are termed *critical chain feeding buffers*. During the project implementation, these buffers will be used to absorb uncertainty or disruptions that may occur in non-critical chain tasks. To determine these buffers, use one-half of the sum of the task durations in the chain of tasks preceding the buffer. No work is assigned to the buffers.

Using the Critical Chain Schedule

The most appropriate application of the Critical Chain Schedule is for a dedicated project team seeking a significant reduction of the project cycle time in a company with an outstanding performance culture. The only job of this team is their project. Equipped with all necessary resources, the team focuses on rapid delivery of value.

CCS can be used for projects where time to market is essential and the organization is trying to reduce development time. The team and the company culture must be willing to use the 50% probable estimates and work to transfer work to the next task in the chain as quickly as possible.

CHOOSING YOUR SCHEDULING TOOLS

Multiple scheduling tools are presented in this chapter that lead to the question, which one or ones are the most appropriate to select and use? Such a decision, of course, depends on your project situation. To help narrow the options, in Table 7.3 we list a set of project situations and indicate how each situation favors the use of the various scheduling tools. Identifying the situations that correspond to your own project is the first step. If the situations described do not describe your particular project situation, brainstorm more situations in addition to those listed, marking how each favors the tools. The tool that has the highest number of marks for identified situations becomes the tool of choice. Note, however, that more than one tool can be used to support any particular project, since some of them complement each other rather than exclude each other. A careful study of the material covered in this chapter will help you determine when this is practical.

Table 7.3: A Summary Comparison of Schedule Development Tools

Situation	Milestone Chart	Network Diagram	Critical Path Method (CPM)	Critical Chain Schedule
Small and simple projects		✓		
Short training time	✓	✓		
Focus on important events	✓			
Increase goal orientation	✓			
Large, complex, and cross-functional projects		✓	✓	✓
Focus on top-priority activities			✓	✓
Strong interface coordination needed		✓	✓	✓
Very fast schedule		✓		✓
Short-term outlook schedule in large projects		✓		
Use of templates desired	✓		✓	✓

References

1. Wheelwright, S.C. and Clark, K.B. (2011). *Revolution Product Development: Quantum Leaps in Speed, Efficiency, and Quality*. New York: Free Press.

2. Steyn, H. (2009). An investigation into the fundamentals of critical chain project scheduling. *International Journal of Project Management*. 19 (6): 363–369.

8

COST PLANNING

Cost is a key factor in project planning, tracking, and statusing. Cost metrics are often the most watched on a project, and delivering on budget is a critical success factor.

Cost planning is the process by which the organizational goals are translated into a plan that specifies the allocated resources, the selected estimating methods, and the desired schedule for achieving the goals of the project. The cost-planning process culminates in the establishment of a project budget that represents the organization's investment in the project.

Project cost estimating can be accomplished in various ways as dictated by the needs of the project and the policies of the organization. Since project managers need to be able to respond to and address varying organizational and project situations, they need to have various cost-planning tools in their PM Toolbox. This chapter describes some widely used project cost-planning tools.

COST MANAGEMENT PLAN

The Cost Management Plan consolidates the information needed to effectively plan, estimate, manage, and track project costs. Among other elements, it provides terminology, estimate types, estimating methods, and the variance thresholds for cost management.

To develop a Cost Management Plan, a project manager should have knowledge about the organization's financial policies, financial rate structure, and project staffing policies.

Financial policies inform which types of cost estimates will be used and for what purpose. Staffing policies provide information on outsourcing labor. Associated with staffing policies is knowing the current labor and overhead rates.

Before delving into the Cost Management Plan, it is useful to present some cost definitions.

Project Management Toolbox: Tools and Techniques for the Practicing Project Manager,
Third Edition. Cynthia Snyder Dionisio, and Russ J. Martinelli.
© 2025 John Wiley & Sons, Inc. Published 2025 by John Wiley & Sons, Inc.

Examples of Cost Estimating Definitions

Direct cost. An item of cost, or the aggregate of items, that is identified specifically with the project. These costs, such as labor, materials, and travel, are charged directly to the project.

Indirect cost. The cost of items such as building usage, utilities, management support labor, services, and general supplies is not easily or readily allocable directly to a project. Indirect costs are accrued and charged to overhead accounts, the sum of which is applied as burden.

Fixed cost. A cost that does not vary with usage. An example may be a database server used for testing on a project.

Variable cost. Expenses that vary according to use, such as the number of hours worked by a person assigned to do tasks on a project.

Most probable cost. This is the cost most likely to occur, which is made up of all the itemized known items and a contingency estimate that together invoke a 50% degree of confidence.

Range of accuracy. This is a prediction of the least expected and highest expected cost relative to the most probable cost. Higher quality of estimate, better scope definition, lower project risks, fewer unknowns, and more accurate estimate pricing will lead to a better range of accuracy.

Contingency. An allowance added to an estimate to cover future changes that are likely to occur for unknown causes or unforeseen conditions. Contingency can be determined through statistical analysis of past project costs or from experience in similar projects.

Developing a Cost Management Plan

A Cost Management Plan may be relatively simple, or for large projects it may be quite robust. Most Cost Management Plans identify the cost estimating process. There are different models for a cost estimating process. The one presented below has eight steps.

1. *Establish the purpose of the estimate.* The purpose of the cost estimate will inform the rest of the cost estimate. An estimate that is used for project selection is different than an estimate used to baseline a project. Knowing the purpose will guide you to the appropriate level of accuracy, estimating methods, and contingency guidelines.

2. *Define the scope.* There are three aspects of defining the scope: product scope, project scope, and cost estimate scope. The product and project scope are first identified in the scope statement and WBS. The more detailed the information about the scope, the more detailed the cost estimate can be. The cost estimate scope identifies what is included in the estimate and what is excluded. For example, are indirect costs included? Are capital equipment costs included? Is internal labor included?

3. *Determine the level of accuracy.* The purpose of the estimate and the degree of scope definition influence the level of accuracy for your cost estimate. A common

way of categorizing levels of accuracy is by order of magnitude, budgetary and definitive estimates. As Table 8.1 indicates, they differ in their use, accuracy, and information they require.[1]

1. *Select an estimating method.* There a several methods to estimate costs. Later in this chapter we will review analogous, parametric, and bottom-up methods.
2. *Identify cost variance thresholds.* A variance threshold indicates when you should take action. All projects will have variances. Establishing a variance threshold indicates which variances need to be addressed, and which can just be watched. A common set of variance thresholds is:
   Green. A variance of less than 5%. This indicates a healthy project, from a cost perspective.
   Yellow. A variance between 5% and 10%. This indicates a variances that is heading toward trouble. In these situations you should look to take preventive action so that the variance does not continue to grow.
   Red. A variance greater than 10%. A red variance requires corrective action to bring the performance back in line with the estimate.
3. *Establish reserve.* There are two types of reserve, contingency reserve and management reserve. Contingency reserve is used to address uncertainty and risk. You can establish reserve for individual deliverables if there is a lot of uncertainty, or for the project as a whole. Projects where there is new technology, not a lot of experience, pricing volatility, complexity, and ambiguity will need more reserve than projects with clear, certain, and well-defined scope.
4. *Validate the estimate.* Validating the estimate involves ensuring the estimating data sources are accurate and relevant, reviewing the calculations, and peer reviews.
5. *Create the baseline.* The cost baseline may also be called the budget. It is generally time phased. Therefore, to create a cost baseline, you need to establish a schedule baseline. There is more detail on creating a baseline in an upcoming section.

In addition to the cost estimating process a Cost Management Plan may include information on funding availability. Some projects are subject to periodic funding, such as needing to expend funds within a fiscal year or having to delay work until additional funding is available.

Table 8.1: Levels of Accuracy

	Order of Magnitude	Budget Estimates	Definitive Estimates
Use	Feasibility study, project screening, budgeting and forecasting	Budgeting and forecasting, authorization (partial or full funds)	Authorization (full funds), bids and proposals, change orders
Accuracy	−30%, +50% before contingency	−15%, +30% before contingency	−15%, +15% before contingency
Information Required	Size, capacity, location, completion date, similar projects	Partial design, vendor quotes	Specifications, drawings, execution plan

The plan may include linkages or references to organizational policies associated with financial management. For example, it may reference policies for make-or-buy decisions since purchases impact the cost estimate and project budget. There may be guidance on average rates for labor based on job scale or percentages for indirect costs.

Tips for Cost Planning

Know your user. Ask questions to clarify their needs, item descriptions, and project scope.

Follow the cost-planning process. Don't skip process steps. If the process doesn't work, change it.

Go beyond a "number-cruncher" mentality. Understand the big picture and philosophy of the project and its customer.

Document everything. Include assumptions, references, sources, scope exclusions, and so on.

Leave an audit trail. Audits enhance quality of the estimate and demonstrate that a process was followed.

Document changes. The estimate you have originally planned for is almost certain to change. Record the change and maintain document version control.

Create involvement and buy-in. Include experts from each performing functional department in estimate preparation; after all, they have to live with the estimate.

Using the Cost Management Plan

While any project can find value in using a Cost Management Plan, large projects benefit most. Consistency and discipline in cost estimating and management is of vital importance to these users.

Developing a cost estimate for a large project is a significant time commitment. When large projects that are also complex and resource-intensive, heavy involvement of experts from various functions—technical, financial, accounting, for example—is typical, often resulting in hundreds of resource hours required to construct a quality cost estimate. For these projects the Cost Management Plan provides a roadmap to guide the process.

ANALOGOUS ESTIMATE

An Analogous Estimate is the derivation of a project cost estimate based on the actual cost of a previous project or projects of similar size, complexity, and scope. The estimators may use historical data, or rules of thumb that are modified to account for any differences between the estimated project and the analogous project(s). An example of the Analogous Estimate is illustrated in Table 8.2.

	1	2	3	4	5	6	7
Item	Analog Size (KLOC)	Analog Productivity Factor LOC/Person-Month)	Analog Effort (Person-Month) 2/3	Target Size (KLOC)	Target Productivity Factor (LOC/Person-Month)	Target Effort (Person-Month) 5/6
1	1	100	10	0.8	80	10.0
2	2	50	40	2.5	40	62.5
3	2	200	10	2.5	160	15.6
4	1	100	10	1.0	80	12.5
5	1	50	20	1.0	40	25.0
Totals	7		90	7.8		125.6

Table 8.2: Example Analogous Estimate for a Software Project

Key: KLOC: Thousand Lines of Code; LOC: Lines of Code.

An Analogous Estimate is generally applied when there is a lack of detailed information about the project. Typically, this is the case early in the project life cycle.

Developing an Analogous Estimate

The quality of an Analogous Estimate is highly dependent upon sufficient information about the project scope, historical information about previous projects, resource requirements, and resource rates.

The first step involves identifying who are the end users of the estimate, the purpose of the estimate, estimating format, list of contributors and their roles, and available resources for creating the estimate. This provides an understanding of a project's scope, size, and complexity features. In our example in Table 8.2, the scope of the target project is broken down into five major items (column 1), each with the targeted size (column 5). Each item may be a key feature of the software product. Now we can go to the database of previous projects with similar features to search for projects with similar size (scope). The most appropriate project (or projects) is selected as the analog. The mapping of analogous features to the target project is fairly straightforward because the two projects share a common set of items. Our example has chosen one analogous project with the same five items. Analyzing actual data about the analogous project indicates size and productivity (columns 2 and 3, respectively), as well as the effort for the completion of each of its items in column 4, essentially an analogous cost estimating relationship). Then we transfer the information from the analogous project to the target, adjusting it for analog elements that are not in correspondence with the target project. Specifically, for item 1 in our example, the project team is less experienced and their productivity (column 6) is judged to be 0.8 (judgment factor) of that for the analog project team (column 3). Applying a cost estimation relationship that divides an item size value (column 5) by the productivity factor (column 6) yields an item estimate (column 7) expressed in resource hours. To convert to monetary terms, we can multiply the hours by the resource rates.

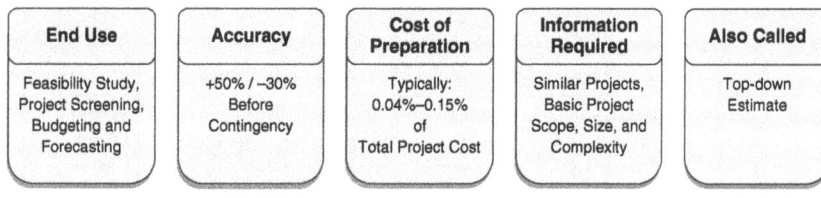

End Use	Accuracy	Cost of Preparation	Information Required	Also Called
Feasibility Study, Project Screening, Budgeting and Forecasting	+50% / –30% Before Contingency	Typically: 0.04%–0.15% of Total Project Cost	Similar Projects, Basic Project Scope, Size, and Complexity	Top-down Estimate

Figure 8.1: Basic Features of Analogous Estimates

A sum of all estimated items is equal to the total project estimate. Crucial in this effort is the ability of the estimators to identify subtle differences in the source and target items and estimate the cost of a target item based on the source item that is analogous. Checking, reviewing, and improving the estimate are the final steps in developing an Analogous Estimate.

The basic features of an analogous estimate is shown in Figure 8.1.

Using an Analogous Estimate

An Analogous Estimate is a tool of choice when there is a lack of detailed information about a project. Typically, this is the case early in the project life cycle. Because other estimating tools have disadvantages of their own as well, an Analogous Estimate can be used in combination with the bottom-up and parametric estimates described in the following sections.

An Analogous Estimate operates on the assumption of a limited amount of information about the target project and summary information about the analogous project. Put together, these two facts mean that just a few hours may be enough for almost any project's analogous estimate. A smaller project would take less than that.

In analogous estimating, an estimator may choose to estimate only the total target project without breaking it down into items as we have done. He or she may judge, for example, that the target project may take twice the resource hours as the analogous one. This judgmental factor of 2 would then be multiplied by the resources deployed in the analogous project to obtain the estimate for the new target project.

PARAMETRIC ESTIMATE

A Parametric Estimate uses mathematical models to relate cost to one or more physical or performance characteristics (parameters) of a project that is being estimated. Typically, the models provide a cost estimation relationship that relates cost of the project being estimated to its physical or performance parameters (also called cost drivers), such as production capacity, size, volume, weight, power requirements, and so forth.

Determining the estimate for a new power plant may be as simple as multiplying two parameters—the number of kilowatts of a new power plant, for example, by the anticipated dollars per kilowatt. Or it may be very complex, such as estimating the cost of a new software development project that requires 32 parameters be comprehended in a cost estimate algorithm.

Parametric cost estimating tends to be faster and less resource-consuming than bottom-up estimating. Focused on the need to establish good cost estimation relationships that properly relate project cost and cost-driving parameters, parametric estimates put a focus on cost-driving parameters, disregarding what is less important. This concentration on cost-driving parameters—coupled with greater speed and lower resource consumption—enables parametric estimates to be applied in estimating situations in which detailed, bottom-up estimates are neither practical nor possible.

Developing Parametric Estimates

Prior to developing a Parametric Estimate, a project manager should have knowledge about the project scope, the key cost drivers, cost estimating relationships (CER) to use for comparison, and historical information from like-projects.

An example of a cost estimating relationships might include hourly labor cost, quantities to be produced, and unit production rates as shown below.

$$\text{Labor cost} = \text{Quantity} \times \left(\text{hrs}\big/\text{unit} \right) \times \left(\$\big/\text{hr} \right)$$

$$\text{Labor cost} = 200\,\text{articles} \times \left(5\text{hrs}\big/\text{article} \right) \times \left(\$80\big/\text{hr} \right)$$

$$\text{Labor cost} = \$80,000$$

If there is enough historical data, the costs are up to date and normalized for quantity, developing a parametric estimate can be very quick. As shown in Figure 8.2, the simplest cost estimate relationships are as simple as a dollar per square foot relationship, a linear relationship of the form $y = ax$, where y is the estimated project cost (dependent variable) that is a function of x, the area in square feet (parameter or cost driver), and a is the parameter based on historical cost data relating to the cost driver. If, for example, this type of cost estimate relationship is used for rough order of magnitude cost estimating for a new home, and assuming that a number of homes between 1800 and 2300 square feet had costs of $500 per square foot, then the corresponding cost estimate relationship can be expressed as $y = 250(x)$.

This simple linear model assumes that there is such a relationship between the independent variable (cost driver) and project cost so that as the independent variable changes by one unit, the cost changes by some relatively constant number of dollars. Clearly, the assumption in applying a cost estimating relationship is that future projects will be performed as past projects.

The above example assumes that you already have a validated cost estimating relationship model based on historical information that has been normalized and kept up to date. If this is not the case, you may need to build your own model. While we won't go into detail on how to accomplish this, Figure 8.3 shows the steps needed to build a CER model.[2]

What if, as is often the case, the future project that is being estimated differs from the past projects in some details? This can be resolved by cost estimating relationship stratification and cost adjustments. Through stratification the historical database is divided into layers, each layer representing a "family" of data points similar to each other

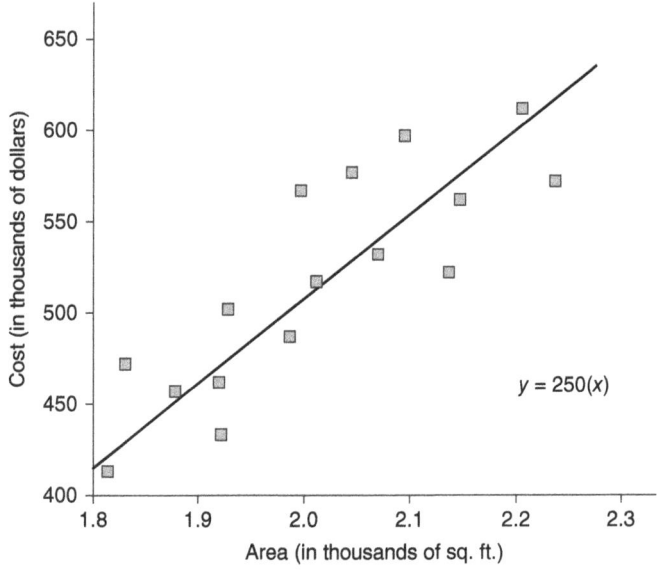

Figure 8.2: Sample Cost Estimate Relationship for Parametric Estimates

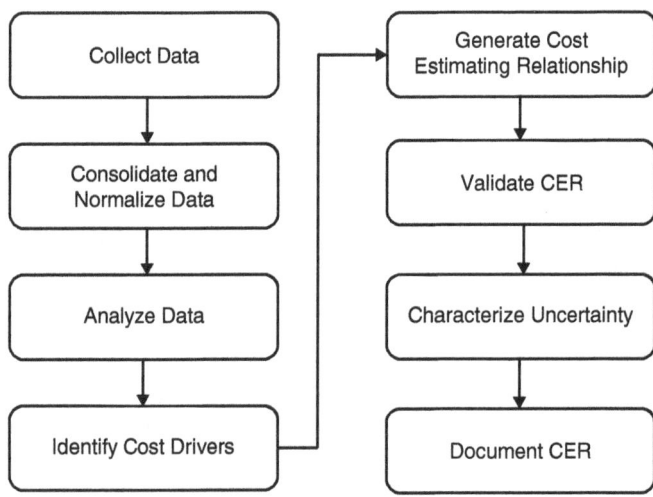

Figure 8.3: Building a Cost Estimating Relationship Model

in some respect. Then, a separate curve for each family is fitted. For example, six data points in Figure 8.2 have higher costs than the other nine. A close look reveals that these six data points are for luxurious homes with features such as central vacuum cleaner, surround sound audio system, stainless steel appliances, marble countertops, hardwood floor, stucco work, and so forth, while the remaining nine were ordinary homes with much simpler and less expensive features. Logically, we could stratify our database into two families of homes and fit curves through each of the two subsets of the database, thus obtaining two cost estimating relationships, as illustrated in Figure 8.4. If we have a square footage of the home being estimated (for example, 2000 sq. ft.), we can easily determine that the parametrically estimated cost of either the luxurious ($186k) or the ordinary home ($173k).

Figure 8.5 shows the basic features of Parametric Estimates for reference when considering the use of such a method.

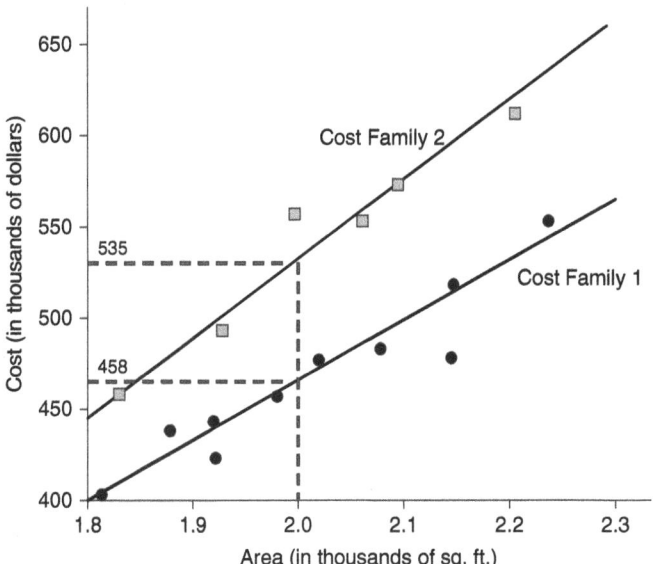

Figure 8.4: Stratified Cost Estimating Relationships

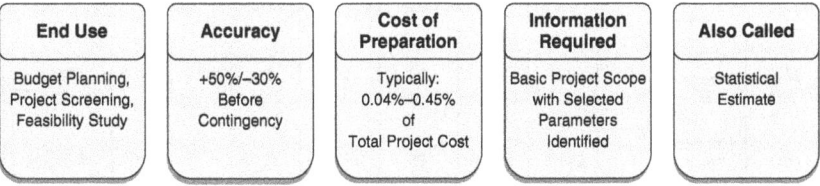

End Use	Accuracy	Cost of Preparation	Information Required	Also Called
Budget Planning, Project Screening, Feasibility Study	+50%/–30% Before Contingency	Typically: 0.04%–0.45% of Total Project Cost	Basic Project Scope with Selected Parameters Identified	Statistical Estimate

Figure 8.5: Basic features of Parametric Estimates

Using Parametric Estimates

Parametric Estimates are most often used in the project definition stage as well as in the early design stages when insufficient information is available to develop a Bottom-up Estimate. Considering that cost estimation relationships typically relate project cost to high-level measurement of capacity or performance, it is exactly this information that is available early in the project cycle. Naturally, such summary information makes parametric estimates very appropriate for calculating comparative cost assessments of alternate project approaches and providing a cross-check to other estimating tools, but not for developing a detailed competitive cost proposal. To be used for such purposes, the Parametric Estimate must be based on accurate historical information, quantifiable parameters, and a scalable model (applicable in both small and large projects). This makes them easy to use, defensible, and repeatable.

The most difficult and time-consuming part of parametric estimating is the model development, including database development and formulation of the cost estimating relationship. Depending on the complexity of the database, it may take anywhere from tens of hours to hundreds of hours to develop the database and cost estimating relationship. Once that's done, actual project estimation can be accomplished in minutes or hours.

BOTTOM-UP ESTIMATE

A Bottom-up Estimate relies on estimating the cost of individual work items and then aggregating them to obtain a total project cost. Typically, an in-depth analysis of all project tasks, components, and processes is performed to estimate requirements for the items. The application of labor rates, material prices, and overhead to the requirements turns the estimate into monetary units. Figure 8.6 is a generic version of the

PROJECT BUDGET ESTIMATE

Project Name: Longfellow   Estimator: Williams   Date: August 5

1	2	3	4	5	6	7	8	9	10	11
Code	Item	Quan	Labor				Over-Head (25%)	Materials		Total $ 7+8+10
			Unit Hours	Total Hours	Rate $/Hour	Amount (5) X (6)		Unit Price	Amnt	
3210	First Article	10	0.5	5	60	300	75	45	$450	$825
010	Project Total	1	291.5	291.5	65	18947.5	4737		$900	$24,584

Figure 8.6: An example of a Bottom-Up Estimate

Bottom-up Estimates for simpler projects, but they can be used to estimate both simple and complex projects.

Typically, a Bottom-up Estimate is developed just before project execution and provides highly accurate estimates. It is what the cost baseline is based on.

Developing Bottom-up Estimates

To develop a Bottom-up Estimate, a project manager needs to have knowledge about the project scope, resources needed and their associated labor rates, material cost required, and the project schedule.

Project scope in the form of a WBS provides a framework to organize an estimate and ensure that all work identified in the project is included and estimated. For this to happen, resource requirements that define types and quantities of resources necessary to complete the work are multiplied by resource rates to obtain a cost estimate. Typically, the rates come from historical records of previous project results, commercial databases, or personal knowledge about team members. Considering that some estimates contain an allowance for cost of financing such as interest charges, which are time-dependent, the durations of activities as defined in the project schedule are an important input.

Prepare the Estimate

To prepare a cost estimate, start with identifying the format, for example, will you organize it by WBS structure, a code of accounts that aligns to organizational accounting codes, or some other method. In our Bottom-up Estimate example, the code in column 1 uses the WBS structure coding. This simplifies the analysis of the project while serving as the basis for cost reporting and cost control. Next determine how you will categorize work, again, you can use the WBS at a control account level, or different types of cost (outsource, material, etc.).

Begin with identifying a work item that is being estimated, then determine its quantity, along with the cost of labor (human resources), overhead, and materials.

We take that approach in our example cost estimate, where a WBS work package is selected as an item. The first work package is called *1ˢᵗ article* (column 2). By repeating this process work package by work package, and then aggregating or totaling the individual costs, we obtain the total project cost.

The *1ˢᵗ article* item requires that the quantity (column 3) of 10 prototypes of a high-tech cable be produced with equipment, tooling, fixtures, and materials that will be used later in the course of the regular production. Should the estimated labor cost for a single unit be shown, or a whole project batch of 10 items? When project tasks are single and nonrepetitive, the question of quantity is irrelevant. When there are multiple identical items, as in our example, the cost for the whole batch needs to be estimated. Accordingly, our cost estimation relationship will multiply half an hour per unit (column 4) by the ten prototypes (column 3), which is a total of 5 resource hours (column 5), by a rate of $60 per resource hour (column 6) to obtain the cost of $300 (column 7) per item.

Columns 4–7 indicate the labor times and cost for each estimated work package item. While we use monetary units to record cost in columns 6 and 7, we do recognize that many a project manager will not do so, but instead will only record labor or resource hours in columns 4 and 5. This is an acceptable practice in many industries, where project managers are not expected to manage dollars but resource hours only. Actually, when labor time estimates are used for estimating future projects, the category of resource hours is much more relevant than cost. With the passage of time, the accuracy of cost is eroded by inflation and other factors, while the resource (labor) estimates should remain valid.

Once the direct labor cost is calculated, you can move to the labor overhead (column 8). There are no hard-and-fast rules since company policies vary widely. While some companies zealously include labor overhead into the estimate, other companies do not factor overhead labor into the estimate at all. Those who do, often have different overhead rates in different parts of the company, and even from one project to another. Very frequently, this rate is based on a cost estimation relationship, calculating it as a percentage of the direct labor cost in column 7. In our example cost estimate, the rate is 25%. Typically, the overhead rate relates to the wages and salaries of employees who are not directly connected with the project, such as supervisors, administrators, and support personnel.

So far, the estimate includes direct and overhead labor cost for an item, in our case a work package. Now, we will estimate a net cost of materials required for the item completion (column 10), using a cost estimate relationship that multiplies the cost per unit (column 9) by the number of units (column 3). Material cost typically comprises costs of components, raw materials, or services for each item. It can include the cost of larger capital equipment as well, which is left out here for the sake of simplification. While our example for unit prices is based on catalog prices of materials, it is also possible to base it on vendor quotations or standard unit costs for stock items.

By adding the costs of all columns, the total cost per item is estimated. Repeating this exercise for each item (i.e., work package) and summing up estimates for all items will lead to a total project cost estimate. If this were a project for an external customer, this would be the time to add profit margin cost to the estimate. Before finalizing the Bottom-up Estimate make sure to spend time checking, reviewing, and improving it.

During the review process, care must be taken to ensure all cost items are included. For an example of the implications of incomplete estimates, see "The Courthouse Disaster."

How the Courthouse Disaster Was Courted

Halfway through its construction, the courthouse project looked like a sure winner for the contractor. The project was on schedule, contract payments were made in a timely manner, and the owner was happy with project performance. Then Greg, the contractor's project manager who also developed the Bottom-up Project Cost Estimate that was the basis for the project contract, left the company. A month later, Pete, the new project manager, determined that the entire project budget had been spent although a lot of work remained. A quick audit commissioned by management revealed the following:

Greg's project cost estimate was never reviewed by peers or managers. A significant monetary loss was to be expected at the end of the project.

When completed a few months later, the courthouse became one of the biggest losers in the company's history, ending $500K over the estimated cost, almost one-third over the original budget. In the postmortem session, the following improvements were adopted for future cost estimating:

All major estimates will be reviewed by peers and management
All major estimates developed under time pressure will be compared to an independent cost estimate (a cost estimate developed by an independent firm).

Using Bottom-up Estimates

Both small and large projects, whether simple or complex, are good candidates to apply Bottom-up Estimates. Typically, the application occurs just before project execution, or even in earlier phases if the required information inputs are available. This generally means that a substantial amount of design work is completed, often exceeding 60%.

For their detailed nature, Bottom-up Estimates are primarily used for cost control budgets, bids/proposals, and change order estimates. The time to develop a Bottom-up Estimate varies with the size and complexity of a project that is being estimated. A 500-resource hour project without materials and equipment may take an hour or two to Bottom-up Estimate. In contrast, a team of estimators may spend thousands of resource hours preparing a bottom-up estimate for a $400M project.

Basic features of the Bottom-up Estimates are summarized in Figure 8.7.

COST BASELINE

The Cost Baseline is a time-phased budget used to measure and monitor cost performance on a project. It is developed by allocating estimated costs to time periods to reflect estimated costs and when they are expected to occur (see Figure 8.8). The Cost

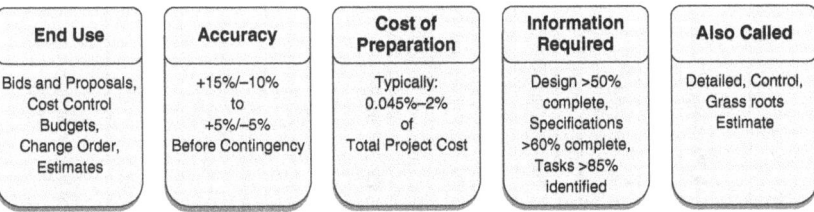

End Use	Accuracy	Cost of Preparation	Information Required	Also Called
Bids and Proposals, Cost Control Budgets, Change Order, Estimates	+15%/−10% to +5%/−5% Before Contingency	Typically: 0.045%–2% of Total Project Cost	Design >50% complete, Specifications >60% complete, Tasks >85% identified	Detailed, Control, Grass roots Estimate

Figure 8.7: Basic Features of Bottom-Up Estimates

Work Packages/Tasks	Item Totals $k	Time Line (in thousands of dollars)												
		FEB	MAR	APR	MAY	JUN	JUL	AUG	SEP	OCT	NOV	DEC	JAN
1.01 Select Concept	12	8	4										
1.02 Design Beta PC	8		1	3	3	1							
1.03 Produce Beta PC	8		1	3	3	1							
1.04 Develop Test Plans	2		1	1									
1.05 Test Beta PC	6					3	3						
2.01 Design Production PC	18						3	6	6	3			
2.02 Outsource Mold Design	16						1	7	7	1			
2.03 Design Tooling	3						5	10	10	5			
2.04 Purchase Tool Machines	16									20	140		
2.05 Manufacture Molds	80									10	10	60	
2.06 Test Molds	8									8			
2.07 Certify PC	18											18	
3.01 Ramp Up	30												30
TOTALS	396	8	7	7	6	5	12	23	23	47	150	78	30

Figure 8.8: A Sample Cost Baseline

Baseline can be depicted as an S-curve that illustrates how labor hours and material are to be expended over the life cycle of a project.

Typically, the baseline is developed in larger projects as part of initial project planning to forecast its cash flow. The Cost Baseline offers benefits as a performance measurement baseline. In this capacity, the baseline is a basis for comparing actual costs (when they occurred) with planned costs (when they were supposed to occur). This provides a way to gauge efficiency and progress, attracting management's attention to any deviations from planned progress and estimated costs.

Cash flow forecasting with a cost baseline informs management or the customer in advance of the funds that must be made available in order to procure resources and use them to sustain project progress.

Developing a Cost Baseline

To develop a Cost Baseline, a project manager needs to have knowledge about the project WBS, the project schedule, and the cost estimate for the project.

A simple definition of cost baselining as the spreading of the cost estimate items over time hints that having a documented cost estimate that includes all cost items is a mandatory starting point. Hopefully, these items can be arranged in alignment with the project WBS, as explained in the section on Bottom-up Estimates. If done so, the knowledge of the project schedule—indicating planned start and expected finish dates for work elements—enables the assignment of the cost to the time period when the cost will be incurred.

Identify Cost Baseline Type and Cost Items

Which types of cost are typically included in a Cost Baseline? Consider including a broad range of cost items, such as:

- Salaries and wages of project personnel (in simplest cases this is the only item to include in in-company projects)
- Overhead expenses
- Payments to contractors
- Payments of vendors' invoices for purchases of equipment
- Materials and supplies
- Interest payable on loans, loan repayments, tax payments, shipping fees, duties, and so on.
- Travel expenses
- Expenses for licenses and permits.

Once the cost items to include are identified, it is time to set criteria for cost baselining.

Set Cost Baseline Criteria

The preparation of a Cost Baseline is essentially an act of establishing the relationship between the cost estimate and time. For this to be possible, there must be clear criteria that determine which project events trigger payments of cost items included in the baseline, and the time intervals between the trigger events and the related payments. For payments to vendors, for instance, the trigger events are usually milestones defined in the contractual terms that stipulate how and when the payments are to be made. For other cost items, such as paying salaries of project team members, labor schedule of their engagement triggers their payment at the end of each month.

Allocate Cost Items to Time Periods

If a Cost Baseline is being developed on the basis of the Bottom-up Estimate, the items can be arranged in line with the WBS, as we have done in Figure 8.8, using work packages from the WBS for a project. When an analogous or Parametric Estimate is being used to construct a Cost Baseline, other methods to arrange the items can be deployed, such as project phases. If the project is externally funded, the customer may mandate the use of its own cost codes.

Column 2 provides cost estimates for the items, which are allocated to certain time periods in the project. Since reporting is on a monthly basis, the time periods shown as months. Item 1.01, Select Concept, is estimated at $12K. Part of the work is scheduled will be carried out in February and the remaining work in March. How much will be allocated to each month hinges on the following factors:

- The project schedule, indicating the planned start and end dates of the item, along with resource histograms specifying resource requirements by time period
- The contractual terms.

Similarly, estimates for the remaining items are spread over their months of execution and entered in the appropriate months.

Sum Estimated Cost by Period

Once all item cost estimates are allocated to specific time periods, the next action is to sum estimated cost by periods. This provides information about incremental expenditures by time periods—that is, expenditures for each month—which will be used in the next step to display the Cost Baseline graphically.

Display the Cost Baseline

The S-curve is a popular way of displaying a Cost Baseline formatted as cumulative expenditures (see Figure 8.9). To calculate cumulative expenditures, add the incremental expenditures for the first period to those of the second period. These are the cumulative expenditures for the first two periods. Add this number to the third period's incremental expenditures to obtain the cumulative expenditures for the first three periods, and continue with this procedure for the remaining periods. When finished, graph the cumulative expenditures (y-axis) over time (x-axis) to develop a Cost Baseline in the form of an S-curve. As in any type of the cost estimate, this is the time to check and review the Cost Baseline.

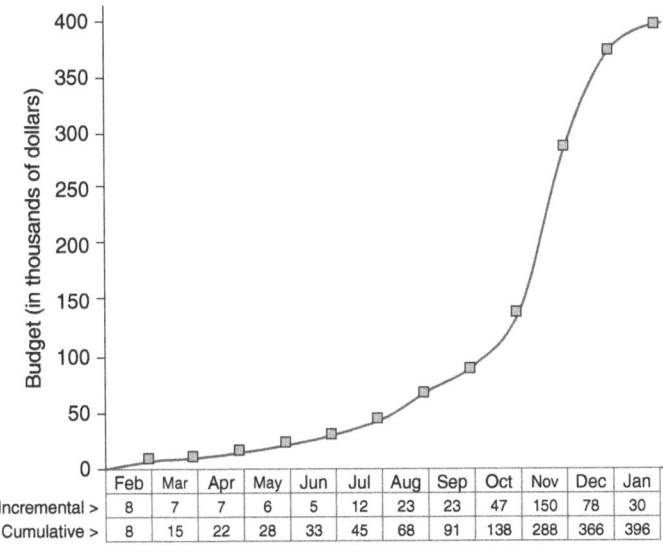

	Feb	Mar	Apr	May	Jun	Jul	Aug	Sep	Oct	Nov	Dec	Jan
Incremental >	8	7	7	6	5	12	23	23	47	150	78	30
Cumulative >	8	15	22	28	33	45	68	91	138	288	366	396

Monthly Budget (in thousands of dollars)

Figure 8.9: A Cost Baseline displayed as an S-curve

Once the project is completed, there is a lot of value in studying how the initial baseline played out over the life of the project, learning the lessons, and using them to improve future Cost Baselines.

Using the Cost Baseline

Smaller projects may choose to forego a cost baseline because the cost of its preparation outweighs the benefits. In contrast, large projects do have a need for the Cost Baseline. Typically, the baseline is developed as part of initial project planning to forecast its cash flow. Considering that the Cost Baseline may be based on an analogous, parametric, or bottom-up cost estimate, sometimes as the estimates evolve and become more accurate, so do the Cost Baselines. They are reissued at regular or irregular intervals and may even constitute part of project reports submitted to senior management or external customers. For details about updating and changing the baseline, see "When Should You Update or Change the Budget?"

As a function of the size and complexity of the project and its schedule, resource requirements, and cost estimate, the time to develop a Cost Baseline may widely vary. The development of a Cost Baseline based on a low-detail Analogous Estimate and summary schedule may consume an hour or two of a skilled project manager's time. On the other hand, an experienced project manager may spend tens of hours constructing a cost baseline based on a detailed Bottom-up Estimate, with hundreds of activities in the schedule.

When Should You Update or Change the Budget?

Dogmatically sticking with the initial Cost Baseline or time-phased budget when there is a need to alter it serves no purpose and is risky. The need for alteration is triggered by several factors, leading to minor (updates) or major revisions (changes) of the baseline. Updates may occur because of factors such as:

Cost estimate evolvement. As a project progresses, more information becomes available, helping to develop more accurate estimates. Such changes in estimates should lead to the update of the baseline.

Project changes. Management of project changes may require new expenditures, which should be added to the baseline. Changes may be due to unforeseen conditions or from customer-generated changes.

Schedule changes. Changes of time-phasing of project activities during the execution stage are frequent and result in inevitable modification of the baseline.

In addition to these updates (minor revisions), there may be times when a major revision of the baseline is necessary. During project implementation, major unplanned schedule, cost, or technical problems may occur. Or, there may be a need to change the project strategy. These, typically, prompt major revision of the project plan, including a major revision of the Cost Baseline. Such changes to the Cost Baseline may happen very rarely, once or twice in the life of a project, if at all. When dealing with updates or changes to the baseline, the key is to manage all modifications and related factors in a proactive rather than a reactive manner, when possible, to maintain control of the project.

Benefits

The lack of an effective Cost Baseline, even if a cost estimate and labor requirements are available, poses a major risk to a project—organizing measurement of performance and cash flow is difficult, if not impossible. Therefore, constructing the baseline offers benefits of using it as a performance measurement baseline. In this capacity, the baseline is a basis for comparing actual costs (when they occurred) with planned costs (when they were supposed to occur). This, then, is a way to gauge efficiency and progress, attracting management's attention to any deviations from planned progress and estimated costs.

Cash flow forecasting is another benefit that an effective baseline provides. It informs management or the customer in advance of the funds that must be made available in order to procure resources and use them to sustain project progress. When properly performing this role in the course of project implementation, the Cost Baseline should be modified to reflect performance and progress to date.

The visual power of displaying a Cost Baseline as an S-curve format is impressive, further strengthening the case of its simplicity and visual nature as a benefit to the project manager and stakeholders.

CHOOSING A COST-PLANNING TOOL

This chapter features five tools with clearly designed purposes. For two tools, the Cost Management Plan and Cost Baseline, the purposes are so distinct that they do not compete with other tools for use in cost planning. While the Cost Management Plan strives to establish a systematic methodology for cost planning, the Cost Baseline aims at providing a time-phased budget.

The remaining three tools may be used in combination or a single tool can be chosen for a particular application. That calls for matching the project situation with the tool that favors the situation. Table 8.3 can provide a project manager guidance on tool selection.

Table 8.3: A Summary Comparison of Cost-Planning Tools					
Situation	Cost Management Plan	Analogous Estimate	Parametric Estimate	Bottom-up Estimate	Cost Baseline
Provide cost-planning methodology	✓				
Show the amount of estimated funds		✓	✓	✓	✓
Show time-phasing of estimated funds					✓
Based on past experience		✓	✓		
Higher accuracy required				✓	✓
Lower accuracy required		✓	✓		
A few hours to prepare		✓			
Medium time to prepare			✓		
Longer time to prepare		✓	✓		✓
Need estimate for project screening, forecasting			✓	✓	
Need estimate for budget authorization				✓	
Make decisions very early in project life cycle		✓			

References

1. Kerzner, H. (2022). *Project Management: A Systems Approach to Planning, Scheduling, and Controlling*. Hoboken, NJ: John Wiley and Sons.
2. Naval Center for Cost Analysis *Cost Estimating Relationship Development Handbook* (2016).

9

RESOURCE PLANNING

Projects depend on resources, both team resources and physical resources. Most large projects do not have all the necessary resources internally; thus, they must procure them. The larger and more complex and project is, the more likely you are to engage in procuring resources from outside the organization. Therefore, to plan and manage your schedule and costs effectively, you need to plan and manage your resources and procurements effectively.

In this chapter we will identify aspects of resources and procurements that need to be analyzed and documented in various plans. These plans support estimates for cost and schedule estimates. The estimates support developing baselines that you will use to compare actual results with planned results.

RESOURCE MANAGEMENT PLAN

A Resource Management Plan provides guidance for both team and physical resources. It documents the expectations for acquiring and assigning team resources along with information on the type of physical resources needed and how they will be acquired and managed.

Team Resources

Team members are the most important driver of project success. Without the right skill level, number, and availability of resource your project will be challenged and perhaps even fail. Once your scope is defined you can start to identify the roles or positions needed on your project. As more detail is provided you will be able to elaborate this information by including the number of resources needed for each role. By the time you are ready to baseline your plan you should also be able to identify the required skill level of the resources.

This information can be documented in a tabular format, like the one shown in Table 9.1.

Project Management Toolbox: Tools and Techniques for the Practicing Project Manager, Third Edition. Cynthia Snyder Dionisio, and Russ J. Martinelli.
© 2025 John Wiley & Sons, Inc. Published 2025 by John Wiley & Sons, Inc.

Table 9.1: Team Resource Requirements			
Role	**Number**	**Qualifications**	**Competencies**
Apprentice electrician	8	High school diploma. 2 years' experience	Wiring installation, system installation
Journeyman electrician	12	Licensed electrician. 5 years' commercial electrician experience	Install, maintain, and repair electrical systems in commercial buildings
Master electrician	3	Electrical contractor license. 10 years' commercial electrician experience	Training, managing, trouble shooting, complex problem solving

In addition to the information in the Team Resource Requirements table, you will want to document the following information to establish clear expectations[1]:

Role. The role of job title along with a brief description.

Authority. The decision-making and approval levels, such as conflict management, prioritization, and alternative selection.

Responsibility. The activities that each role carries out, such as job duties, processes, and hand-offs to other roles.

For projects with a large number of resources you may want to include an organizational chart for the project so there is clarity around reporting structure and the interfaces between various functions.

Long-term projects or projects that require hiring or leasing team members have some additional resource requirements. For example, you may need to develop a training plan to develop new skills for resources, or to educate them on the context and technology for the project.

When your resources are not expected to be part of the project from inception to completion, you may need to establish a process for onboarding new team members. As people leave you will want to have a checklist for releasing them. This can include returning any materials, equipment, or supplies, along with transferring knowledge to the appropriate people.

Physical Resources

Like team resources, the information regarding physical resources is progressively elaborated. You may start with a structured list of resources broken out by category. Figure 9.1 shows an example of a Resource Breakdown Structure for an electrical part of a project.

Throughout the planning process you will get clarity on the amount and grade of resources you need. This can be entered into the Resource Breakdown Structure or if you do similar projects on a regular basis, you may have a checklist of materials.

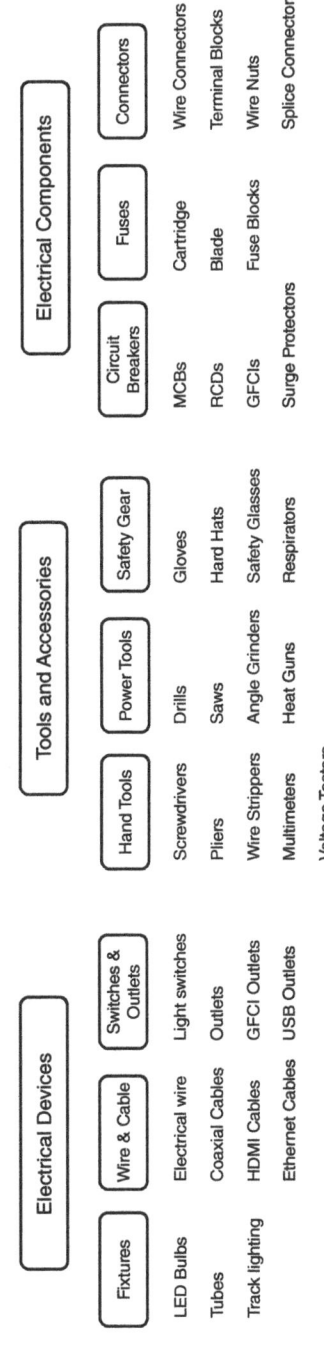

Figure 9.1: Resource Breakdown Structure

Your Resource Management Plan should indicate how physical resources will be acquired. This can include buying, leasing, renting, or drawing from inventory. In situations where you are not drawing from inventory, the Resource Management Plan, the Procurement Management Plan, and a Procurement Strategy should all align.

You will also document how physical resources will be managed. This may include inventory management, supply chain management, and logistics information.

Using

Your Resource Management Plan reflects the planning needs of your project. For large complex projects with many resources, you will have a robust plan. This plan will be utilized throughout the project as team resources come and go, and as physical resources are acquired and expended.

Smaller projects may use the Resource Management Plan primarily for planning and communication purposes. If the project has a dedicated team and does not require any procurements, a few pages that reflect resource needs is sufficient.

Benefits

The Resource Management Plan is part of an integrated Project Management Plan. Having a separate subsidiary plan that is dedicated to team member and physical resources provides a structure to help organize resource information.

The section on the team resources provides the team structure, roles, and skills to ensure there are no skill gaps. It confirms that all the team processes, such as onboarding, training, and knowledge transfer, are addressed and tailored appropriately.

Structuring and organizing the physical resources is of paramount importance for a large project. The Resource Management Plan provides an overarching view of the physical resources needed for the project, and how they will be acquired and managed.

RESPONSIBILITY MATRIX

A Responsibility Matrix (sometimes known as a Responsibility Assignment Matrix or RAM) provides a clear view of who is involved with the project, what they are responsible for, and how they will participate. It is an effective tool for establishing this *"who," "what,"* and *"how"* relationship between the project components and the project players.

Developing a Responsibility Matrix

A Responsibility Matrix aligns with the project Work Breakdown Structure (WBS). At the start of a project, it may reflect the work done at Level 1 or Level 2 of the WBS. It uses the information from the team resources requirements to determine who will participate with the various deliverables from the WBS and how they will participate. It is common during the early stage of a project that some project players are not yet identified by name. When this is the case, use a role designation, such as Architect, instead of a person's name.

Task	PM	Architect	General Contractor	Foundation Contractor	Carpenter	Finish Carpenter	Utility Contractor
Administration							
Architectural drawings							
Substructure							
Structure							
Finish work							
Utilities							
Permits							
Project management							

Figure 9.2: Initial Responsibility Matrix Structure

At this point, the basic structure of the Responsibility Matrix is created, see Figure 9.2. As shown, top-level elements from the WBS are listed in the left-hand column of the matrix. The team members identified (or roles) are in listed along the top row.

Select Responsibility Designations

In order for the matrix to be effective, it must accurately reflect people's expectations and responsibilities. This is accomplished by using different participation types that designate the appropriate level of responsibility for project outcomes.

Many variations of participation designators exist. One of the most common is called a RACI chart where:

"R" means the person *responsible* for performing the work
"A" means the person *accountable* for the work's completion
"C" means a person *consulted* for information about the work
"I" means a person who is informed about the work.

While there may be many people may be responsible, consulted, or informed when working on a deliverable, there is always one, and only one person who is *accountable* for the outcome. That person is usually a team lead, or a subject matter expert. While it may be tempting to note that ultimately the project manager is accountable for everything about the project, this is the tool the project manager uses to delegate accountability to those who are most qualified to manage the work.

Assign Responsibilities

The final step in developing a Responsibility Matrix involves assigning the appropriate responsibility designators to the team members. Figure 9.3 illustrates a completed Responsibility Matrix for a construction project.

Task	David P	Avi R.	Victor P.	Aaron M.	Jorge M.	Chip K.	Siva G.
Administration	AR						
Architectural drawings	CI	AR	CI	I	I	I	I
Substructure	I		AR	R			
Structure			AR	R			
Finish work			CI		R	AR	
Utilities			I				AR
Permits	I	R	AR	R	R		R
Project management	AR	R	R	R	R	R	R

Figure 9.3: Completed Project Responsibility Matrix

Note that there are several instances where the designation "AR" is used. This denotes that the person accountable for the outcome is also performing the work. In contrast, there are several instances where a person is responsible for doing the work, but not accountable.

Using the Responsibility Matrix

A Responsibility Matrix is useful in the beginning of the project. It helps to match up the key deliverables with resources. If desired, as the work is progressively elaborated the Responsibility Matrix can be elaborated as well. However, when a detailed schedule is developed, the schedule fills the need to determine who is assigned and working on the lower-level activities. The information on the Responsibility Matrix is still valid, but it is kept at the deliverable level rather than the task level.

The Responsibility Matrix is a living project artifact. Project team members and responsibilities will change during the course of a project, so too must the Responsibility Matrix. Remember to update the matrix whenever major responsibilities for project outcomes shift.

While the Responsibility Matrix is shown here, there are other options. Some project managers include an "S" to represent the person who signs off or approves a deliverable. Other people use "C" to indicate who is creating a deliverable, rather than consulting on a deliverable. The designations are less important than the clarity that the matrix brings.

Benefits

Using a Responsibility Matrix early in a project brings clarity about who is account-able for key deliverables. It brings together two critical activities—identifying the key deliverables and formation of the core project team. The Responsibility Matrix gives the project manager a tool for reducing this uncertainty and ambiguity that are frequently present at the start of a project.

As stated earlier, a project manager has to delegate accountability and responsibility of outcomes to other team members associated with the project. By documenting delegated responsibility through the Responsibility Matrix, the necessary delegation is established and communicated to primary stakeholders of the project and top management within an organization.

This removes possible implied assumptions both on the part of the project team members and the stakeholders and replaces it with explicit direction. This can be particularly useful *between* organizations when multiple companies are involved in a collaborative project agreement.

PROCUREMENT MANAGEMENT PLAN

Most large projects have one or more procurements. Procurements can include goods, services, equipment, material, supplies, and manpower. Anytime you introduce procurements into a project, you are increasing the complexity. For example, there are legal considerations based on contracts, integration considerations for schedule and reporting, and the extra work associated with planning, conducting, and managing the contracting process.

Project managers rarely have the authority to enter into a legal contract on behalf of the company. Therefore, it is important to establish a good working relationship with the people that do have that authority. This is usually someone in purchasing, legal, or compliance departments.

A procurement management plan is used to address procurements for the project as a whole, rather than individual contracts and purchases. While the plan guides the procurement activities associated with the project, it must be developed with the input and approval of people with the authority to conduct and manage procurements.

Developing a Procurement Management Plan

The Procurement Management Plan provides the overarching guidance for all project procurements. When there are only a few procurements the information in Procurement Management Plan can include information on specific procurements as well. However, when there are several procurements, it is a good practice to address information on individual procurements separately in a Procurement Strategy document.

There are many activities necessary to procure items. These activities should be defined in the Procurement Management Plan and incorporated into the schedule. Table 9.2 describes examples of procurement activities.

Table 9.2: Example Procurement Activities	
Documenting terms and conditions	Terms and Conditions are the legally binding rules for the contract. They include the rights and obligations for each party, information on delivery and payment, governing law, and confidentiality. They also outline dispute resolution agreements, renewal, and termination, and how amendments and modifications can be made
Preparing a Statement of Work (SOW)	A Statement of Work describes the tasks, deliverables, timelines, and other essential details that a contractor will undertake for a client
Managing the bid process	Managing the procurement process includes advertising the opportunity, preparing bid documents, such as Request for Proposal (RFP) or Request for Quotation (RFQ), evaluating proposals, and selecting the winning proposal. Once the winning bidder is selected, there are often negotiations that need to take place prior to finalizing the contract
Defining roles, responsibilities, and authority for the procurement	Due to the legal nature of contracts, and the often large sums of money involved, there needs to be a clear understanding of who is involved, what their responsibilities are, and what level of authority they have. For example, a contracting officer may be accountable for documenting the Terms and Conditions. The project manager will likely create the first draft of the Statement of Work and collaborate with the contracting officer and technical subject matter experts to finalize it. Usually, only someone from the procurement, purchasing, or legal department has the authority to commit the company to a contract
Identifying jurisdiction	Identifying the legal jurisdiction is important, especially when contractors are in different states or different countries. In the situation where contractors are in different countries, you should identify the monetary units and/or expected exchange rates
Administering payments	Managing the contract entails administering payments, auditing contractors, and managing any claims per the terms outlined in the Terms and Conditions of the contract.

To provide consistency in how contractor's work will be integrated into project work, it is useful to establish guidelines in the Procurement Management Plan. Integration considerations include:

- *Scope.* The process to ensure that contracted scope integrates with inhouse scope and that there are no gaps or redundancies.
- *Schedule.* Documenting procurement activities and milestones into the master schedule. Additionally, identifying delivery dates and other contractor milestones into the master schedule.
- *Cost.* Establishing payment guidelines for procurements, whether it is based on deliveries, periodically, or some other method. If incentive and award fees are expected, processes for assessing the fee and allocating budget for it should be included.
- *Risk.* Identify risks associated with outsourcing work. Determine which type of contractor risks should be documented in the project risk register and which the contractor can manage.
- *Reporting.* Establish a reporting format for contractors and expectations for the timing of reports.
- *Documentation.* Define the types of documentation that contractors should develop and transfer to the project manager.

For projects with multiple procurements, the Procurement Management Plan may include a high-level milestone schedule or roadmap that shows each of the major procurement milestones for each contract.

The Procurement Management Plan should be analyzed in context with other planning documents, such as the schedule, budget, WBS, and requirements to ensure there is alignment.

Using a Procurement Management Plan

The Procurement Management Plan is referenced anytime there is a procurement. Thus, it may be used at the very beginning, during, and at the end of the project. Much of the guidance supports the bid and source selection process; however, the majority of the contract administration occurs during the life of the contract.

During the life of the contract, the Terms and Conditions will be applied, technical progress will be assessed, and payments will be administered. For long-term contracts it is common to conduct periodic site visits. The site visits augment the status reports to ensure there are no misunderstandings or surprises. Site visits may include a procurement audit. Any deficits or concerns should be addressed by the buyer and seller project managers as well as the contracting officer.

Despite the best intentions, there are times when the buyer and the seller disagree on deliverables, change requests, or payments. When this occurs a claim is filed. A claim is formal documentation of a disagreement. If a claim cannot be resolved, it becomes a dispute. Claims and disputes are handled according to the verbiage in the contract. It is always preferable to handle the claim through negotiation. However, if that is not possible, most contracts define a process for Alternative Dispute Resolution (ADR).

When the contract is complete and all payments have been made, the contract is closed. It is usually the contracting officer that has the authority to close a contract. The Procurement Management Plan describes the contract closure process from the project perspective, such as the transfer of all intellectual property, complete documentation, and any other support needed for the goods or services provided.

Benefits

Because procurements entail contractual obligations, clearly documented guidance on how to enter into, administer, and close contracts is imperative. The process of planning, preparing, and entering into a contract can be quite complicated, especially for large contracts and/or highly technical work. The Procurement Management Plan incorporates corporate policies as well as project-specific guidance to clarify and streamline the process. This allows the project manager to focus on the technical work rather than figuring out process and procedural work.

Projects with multiple procurements find a Procurement Management Plan particularly useful, especially when there is a milestone schedule or roadmap that identifies the dates for significant procurement activities.

PROCUREMENT STRATEGY

Unlike a Procurement Management Plan, a Procurement Strategy is used for individual procurements. The Procurement Management Plan identifies the procurement activities, such as developing a Statement of Work, identifying bidders, and establishing source selection criteria. The Procurement Strategy applies those activities to a specific procurement. Typical contents for a Procurement Strategy include the contract lifecycle, contract type, fees, and source selection criteria.[2]

Developing a Procurement Strategy

Each procurement goes through a lifecycle with a set of phases from inception to completion. Figure 9.4 shows a generic procurement lifecycle.

In the Preparation phase the project manager and the contracting officer work together to create, organize, and publish all the documentation needed to conduct the bid process. Typical work includes:

- *Identify vendors.* Often purchasing agents will have a list of pre-qualified vendors, especially if the organization frequently outsources the type of work being contracted. If there are no pre-qualified vendors, or if the list is not sufficient, then the research will need to be done to find qualified vendors.
- *Develop statement of work.* Developing the technical statement of work is usually coordinated by the project manager. Subject matter experts provide the knowledge for the statement of work and the project manager with the aid of the contracting officer may finalize it.
- *Determine contract type.* There are three categories of contracts: fixed-price, cost-reimbursable, and time and materials. Within these categories there are different types of contracts to suit the needs of the buyer and the seller. Part of determining the contract type is determining if there will be incentive fees or award fees as part of the contract, and if so, what the criteria for the fees or awards will be.
- *Establish source selection criteria.* Establishing source selection criteria prior to receiving the proposals makes the selection process more transparent and objective. Selection criteria can include cost, delivery schedule, technical approach, experience, and other criteria applicable to the contract.
- *Develop bid documents.* Bid documents may take the form of an Invitation for Bid (IFB), Request for Proposal (RFP), Request for Quotation (RFQ), among others. For procurements with clear requirements that don't include innovation or new technology, an RFQ is generally sufficient. The more uncertainty associated with the bid, the more likely you are to present an RFP to assess the vendors proposed solution.

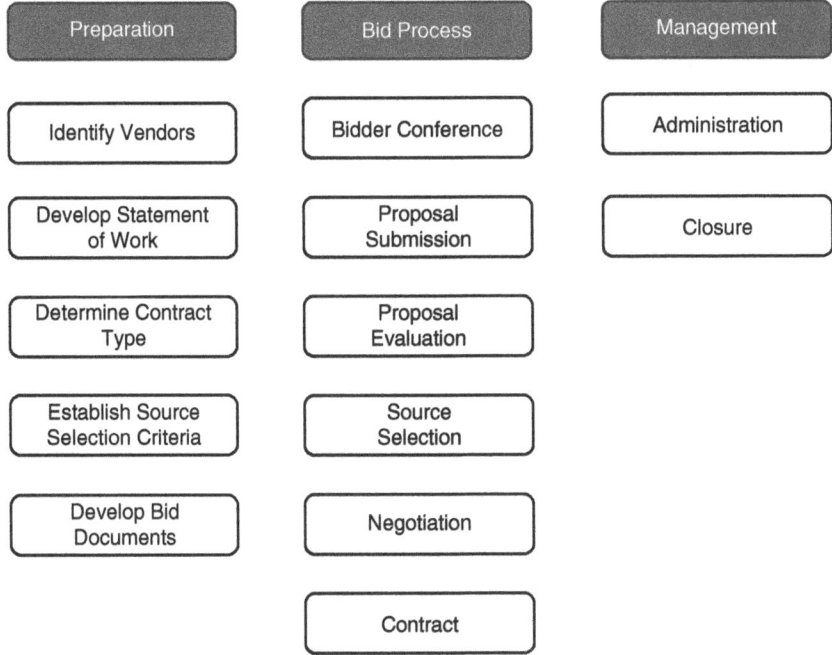

Figure 9.4: Generic Procurement Lifecycle

The Bid Process phase begins once the bid documents have been sent out and culminates in a signed contract. The steps involved are:

- *Bidder conference*. Bidder conferences aren't necessary on all procurements. When they are used, they are an opportunity for vendors to ask questions and get clarification. If the buyer has new information, it is imparted at a bidder conference. It is important that all bidders receive the same information so there is no favoritism shown or implied.
- *Proposal submission*. During this time the sellers prepare their responses to the bid documents.
- *Proposal evaluation*. Depending on the type of work that is being contracted, there may be several subject matter experts in addition to the project manager and contracting officer that evaluate the proposals. Some procurements have a multi-step evaluation process. The first step may be evaluating the proposals for completeness and compliance with the bid directions. The next step could be a technical evaluation. Another step may be a cost analysis.
- *Source selection*. Once the bids are evaluated from a compliance, technical, and cost perspective the remaining bids can be scored using the source selection criteria that was established in the Preparation phase. This may take the form of a decision matrix that is either weighted or unweighted. The result of the scoring process may show a clear winner, or may reduce the number of viable bids to enter into a negotiation process.

- *Negotiation.* During negotiation all aspects of the SOW and the contract are negotiated. This can include delivery dates, payment schedule, terms and conditions, place of delivery, and so forth.
- *Contract.* Once the bid winner is selected, the work of drawing up the contract begins. Getting a signed contract for large procurements can take some time. It is not uncommon to use a Letter Contract to begin work, or to use a Time & Materials contract until the final contract is signed.

The final phase in the procurement lifecycle is management. In this phase the project manager integrates the work of the contractor with the project team and other contractors to further the work of the project. The contracting officers manage the payment and any disputes that arise. When the work is complete and approved by the project manager, the contracting officer closes out the contract.

Using a Procurement Strategy

The procurement lifecycle phases and activities for each procurement should be integrated into the WBS and the schedule. Large procurements are like their own project. They have defined scope, schedule, cost, requirements, risks, stakeholders, and resources. The lifecycle for this type of procurement may have its own schedule that is incorporated into an integrated master schedule.

For smaller procurements, or when there are only a few procurements in a project, the information in the procurement strategy can be integrated into the Procurement Management Plan.

Benefits

The procurement strategy provides the framework for the procurement from identifying the vendors through issuing the contract. Having a defined strategy for each purchase helps organize and prioritize all the procurement work needed to effectively manage one or more procurements on a project. Establishing a consistent lifecycle with clear deliverables and criteria for advancing to the next phase reduces uncertainty and sets clear expectations for all stakeholders.

Organizations that utilize contract work on a consistent basis often have templates and historical data that can help reduce the amount of work associated with each procurement.

References

1. Dionisio, C.S. (2023). *A Project Manager's Book of Templates*. Hoboken, NJ: John Wiley and Sons.
2. Dionisio, C.S. (2023). *A Project Manager's Book of Templates*. Hoboken, NJ: John Wiley and Sons.

10

RISK PLANNING

Project risk is associated with uncertainty. Uncertainty can lead to positive or negative outcomes. By understanding what creates uncertainty and risk on a project, we are able to manage it proactively. However, without good risk management practices and tools, we can find ourselves in crisis management as problem after problem presents itself, forcing a team to constantly react to the problem of the day. As one well-known author stated, "If you don't actively attack the risks, the risks will actively attack you."[1] Risk management is a preventive practice that allows the project manager to identify potential problems *before* they occur and put corrective action in place to avoid or lessen the impact of the risk.

Understanding the level of risk associated with a project is important for several reasons. First, by knowing the level of risk associated with a project, you understand the amount of schedule and budget reserve (risk reserve) needed to protect the project from uncertainty. Second, risk management is a focusing mechanism which provides guidance as to where critical project resources are needed—the highest risk events require adequate resources to address them. Finally, good risk management practices enable informed risk-based decision making. Having knowledge of the potential downside or risk of a particular decision, as well as the facts driving the decision, improves the decision process by allowing the project manager and team to weigh potential alternatives, or tradeoffs, to optimize the reward/risk ratio.[2]

The tools presented in this chapter are instrumental in identifying risks to the project, assessing their potential impact, and developing actions to reduce or eliminate them. Although there is certainly no shortage of project risk management tools to include in a PM Toolbox, we have included the tools that we see most widely used in planning and that provide the broadest application to project types and sizes. Advanced risk tools are presented in a later chapter. We begin with the Risk Management Plan.

Project Management Toolbox: Tools and Techniques for the Practicing Project Manager, Third Edition. Cynthia Snyder Dionisio, and Russ J. Martinelli.

RISK MANAGEMENT PLAN

The Risk Management Plan establishes the framework and methodology the project team will use to identify, monitor, and manage the risk associated with a project. Developing a Risk Management Plan at the onset of a project can help to eliminate potential issues from emerging, or at least minimize the impact that they have on the project if they do occur. The Risk Management Plan provides an organized approach for project managers to deal with project uncertainties when they arise and in the best case head them off *before* they impact the project.

Developing a Risk Management Plan

Much of the time projects operate in an environment of uncertainty such as incomplete information and indeterminate outcomes. This is the realm of project risk management. Beyond it lies the region of total uncertainty, with complete absence of information, where nothing is known about outcomes. Figure 10.1 shows a continuum of total certainty (knowns)—risk (known unknowns)—total uncertainty (unknown unknowns).

Risk Management Approach

The Risk Management Plan is a document developed in the beginning of the project that provides a framework for dealing with risk throughout the project's lifecycle. Included in the plan is a general description of the approach used to identify, assess, manage, and monitor project risk events. It should include information such as:

- *Risk management methodology*. Identify and describe approaches, tools, and data sources that may be used to handle risks.
- *Roles and responsibilities*. Define who does what in risk management in the project, from project team members to members of the company's risk management teams.
- *Budgeting and timing*. Specify the budget for risk management for the project, as well as the frequency of the risk management processes.
- *Tools*. Describe which specific methods for qualitative and quantitative risk analysis to use and when to use them.
- *Reporting and monitoring*. Define how risk will be reported and communicated to the project stakeholders, how risk events and triggers will be monitored throughout the project, and how the information will be preserved for purposes of lessons learned.

The risk management approach is fairly well documented and used across most industries, and focuses on the simple cycle of risk identification, assessment, response planning, implementation, and monitoring. Basic risk management terminology is defined in "Basic Risk-Related Definitions."[3]

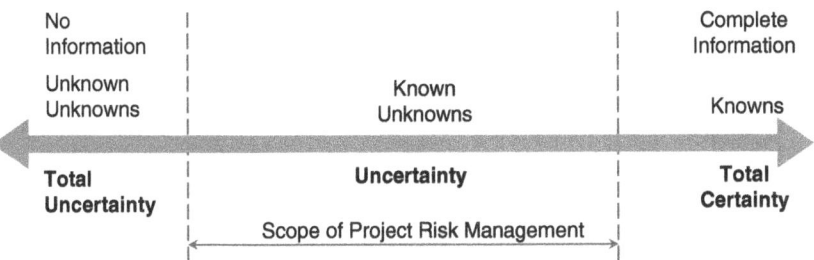

Figure 10.1: The Project Risk Management Continuum

Basic Risk-Related Definitions

Project risk. An uncertain event or condition that can have an impact on a project.

Threat. A risk that will have a negative impact on a project.

Opportunity. A risk that will have a positive impact on a project.

Probability. The likelihood that a risk will occur.

Impact. Severity of its effect on the project objective. Also called risk consequences or amount at stake.

Risk tolerance. The degree of uncertainty an entity is willing to accept.

Risk threshold. The amount of risk exposure an entity is willing to accept before actively addressing a risk.

Contingency reserve. Funds or time incorporated into a baseline to accommodate known risks and risk responses.

Issue. A current condition that may have an impact on a project.

Risk Identification

This section of the plan describes the process for identifying the potential risks that may influence the success of the project. It should describe the methods for how the risks are identified and the format in which the risks are recorded.

Multiple ways for accomplishing this step are available, ranging from engaging the project team in a brainstorming session, to consulting experienced team members, and requesting opinions of experts not associated with the project. Typical methods of identifying risk are expert interviews, reviewing historical information from similar projects, conducting a risk brainstorming meeting, and using more formal techniques such as the Delphi method (see "The Basics of Delphi").

The Basics of Delphi

The Delphi technique is useful in situations where outcomes or trends are uncertain. This is particularly true in situations that have not been encountered previously.

The technique involves soliciting the opinions of a group of experts related to the future situation. It is best if forethought and preparation is exercised to identify specific knowledge gaps relating to the future situation under scrutiny. The knowledge gaps will guide the selection of the right panel of experts and will focus the discussion topics on the right areas.

The technique can be applied informally, where the panel of experts is gathered together to discuss and debate the topics associated with the future scenario, focusing on the knowledge gap areas. Group opinions and ideas are then gathered and summarized. A second or third round of discussion ensues until convergence of opinion begins to occur.

When applied formally, questionnaires are given to each panel expert in two or more rounds. After each round, expert opinions are collected and summarized, and then sent out to each panelist to consider for his or her next response. The process can continue until convergence or a solution arises.[4]

Be acutely aware that complete consensus is highly unlikely and not very useful. Judgment must be applied on the part of the facilitator to gage when convergence (not consensus) has occurred. When used to identify potential risk events, the value of Delphi comes from the *emergence* of ideas, not in the *convergence* of opinion.

When identifying risks, several things have to be taken into account. First, risks vary across the project life cycle. Typically, risk levels tend to be relatively high early in the project because so much is unknown. Similarly, later in the project many of the unknowns are turned into knowns and risk levels are relatively lower. Also, some risks occur only in certain project stages; for example, risks related to project acceptance tests are typically encountered near the end of the project. Sometimes, even assumptions may become a source of risk (see the example titled "Is an Assumption a Risk?").

The dynamic nature of risk makes the identification process iterative. This includes reviewing risks that are identified early in the project as well as emerging risks that are identified during project implementation.

Second, risk events may not strike independently. Rather, they may interact with other risk events, combining into larger risks. Looking for such interactive possibilities is important in risk identification. Finally, since risks come in all types of packages, planners should conduct risk identification in a systematic way so that no stone is left unturned internally in the project and externally in the environment, including management of stakeholders. A huge help in this respect is risk categorization.

Is an Assumption a Risk?

"Is an assumption a risk?" This was a question that a project manager asked in a risk identification meeting. Assumptions are factors that are not entirely known or are uncertain but for planning purposes are considered to be true or certain. For example, a firm launched a project to develop and market a product in a Pacific Rim country. A major assumption was that the country's annual market growth rate would continue to be around 10%. Per its assumption management practice, the firm first documented the assumption by defining it, nominating its owner, and identifying a monitoring metric.[5] Next, the project manager instructed the owner to periodically test the metric in order to ensure that no change of assumption occurred. Seeking to be proactive, the owner defined at which time the assumption becomes a risk (trigger point) and potential risk response actions may be needed.

A few months later the country was hit by a recession and the growth rate turned negative. The project team revisited the assumption and, since the recession was expected to last for some time, decided that the assumption changed into a risk, immediately invoking the Risk Response Plan. So, "Is an assumption a risk?" It is not, rather, it is a source of a potential risk.

Risk Categorization

A good practice when identifying risks is to have risk categories. You may want to categorize risks by their impact on a baseline or project objective, such as scope, cost, time, and quality. There are several frameworks available to categorize risks:

- *PESTLE*. Political, economic, social, ecological, legal, and environmental.
- *TECOP*. Technological, economic, commercial, operational, and political.
- *VUCA*. Volatility, uncertainty, complexity, ambiguity.

The point is that the firm and its projects need a consistent risk categorization schema, suiting its business and culture, which can serve as a framework for a systematic identification and treatment of risks.

Qualitative Risk Assessment

Once risks have been identified, the team will need to determine which risks to actively address, and which to keep on a watch list. The Risk Management Plan provides a structure to assess risks.

Risk Assessment Matrices come in several different formats which project managers can search and adopt for their use. The format shown in Table 10.1 is a Probability and Impact (PxI) matrix. It is by and large the most common.

The matrix is simply a 5×5 matrix with risk probability represented along one axis (in this case the vertical axis), and risk impact along the other axis. For smaller projects

Table 10.1: Sample Probability and Impact Matrix

Probability ↓

VH = 5				
H = 4				
M = 3				
L = 2				
VL = 1				
Impact ➡	VL = 1	L = 2	M = 3	H = 4	VH = 5

Table 10.2: Example Five-level Scale of Risk Probability

	1	2	3	4	5
Scale	Very High	High	Medium	Low	Very Low
Probability	81–100%	61–80%	41–60%	21–40%	1–20%

Table 10.3: Example Five-Level Scale of Risk Impact

	1	2	3	4	5
Scale	Very Low	Low	Medium	High	Very High
Schedule impact	Slight schedule delay	Overall project delay <5%	Overall project delay 5–14%	Overall project delay 15–25%	Overall project delay >25%
Cost impact	Less than 2% overrun	2–5% overrun	5–10% overrun	10–20% overrun	>20% overrun
Scope impact	Minor issue with a single feature	Slight degradation of functionality	Moderate degradation of functionality	Several functions are inoperable	Product is essentially useless

you may want to use a 3×3 matrix to rate high, medium, and low. Some projects get more granular and use a 9×9 matrix.

The next step is to define the scales for which risk probability and risk impact will be assessed. Remember that risk assessment at this stage is qualitative, not quantitative even though numerical values are used to represent qualitative scales. For that reason, the rating scales must be simple and explicit to enable consistency in qualitative assessment for all risk events.

In the example in Table 10.1, we use a five-level scale for Very High (VH), High (H), Medium (M), Low (L), and Very Low (VL). These qualitative values are not sufficient, however. A description of each value has to be documented in the Risk Management Plan for rating consistency. Table 10.2 shows an example of a five-level scale for probability.

In the same fashion, a discrete scale has to be determined for the impact of project risk. This scale has to be based upon specific details of the project. Table 10.3 illustrates

Table 10.4: Sample Probability and Impact Matrix with Severity

Probability ↓	■ High	□ Medium	□ Low		
VH = 5	5	10	15	20	25
H = 4	4	8	12	16	20
M = 3	3	6	9	12	15
L = 2	2	4	6	8	10
VL = 1	1	2	3	4	5
Impact ➡	VL = 1	L = 2	M = 3	H = 4	VH = 5

how a risk impact scale might be constructed based on impacts to schedule, cost, and scope.

The final scale to be defined is the risk severity scale. This is commonly accomplished by multiplying the probability times the impact to develop a risk score, then determining which combination or probability and impact is considered a high risk, medium risk, and low risk. Table 10.4 shows a balanced PxI matrix with an even distribution of high, medium, and low categories. If you have one project objective that is more important than the others, you may choose to show more squares in the high category and fewer in the low category.

If you don't have much data to reliably assess quantitative probabilities, qualitative assessment of risks is sufficient for risk assessment. For projects where sufficiently reliable data is present, you would move onto quantitative risk assessment. Tools used in quantitative risk assessment are described in the chapter on Advanced Risk Management Tools.

Risk Management Roles

While everyone on the project is responsible for identifying, assessing, and addressing risks, there may be some team members with clearly defined accountabilities for risk associated with their roles. These may be documented in a responsibility matrix developed specifically for risk management, or they may be documented in the risk management plan in a section on Roles and Responsibilities.

Some projects have an assigned Risk Manager who manages the entire risk management process from developing the risk management plan and risk report templates and includes performing risk audits.

Technical team leads may be accountable for further decomposing the risk categories for technical risks. They may also be accountable for ensuring that agreed upon risk responses are implemented effectively and in a timely manner.

Funding and Timing

Risk management has costs associated with it. Costs can include bringing in outside help to understand the risk associated with a new technology, funding for risk responses, and contingency funds. The Risk Management Plan should document the estimated funds needed, and how those funds will be managed and tracked.

Risk activities should occur on a regular cadence. Here are some examples:

- A standing agenda item for risk during status meetings
- A full risk reassessment on a quarterly basis
- Monthly Risk Reports
- Bi-annual risk audits

Using the Risk Management Plan

There is no project that cannot benefit from developing and using a Risk Management Plan because all projects contain an element of uncertainty about the future. Small projects typically rely more upon the qualitative assessment of risk, often deciding to handle only a few highest-ranked risk events. Not surprisingly, the dominant mode of risk management planning is informal, as is the periodic re-evaluation of the plan throughout the project.

The Risk Management Plan has more formality and a stronger orientation on quantitative risk assessment and prioritization for larger and more complex projects. Focused on the larger number of highest-ranked risk events, larger projects also tend to do more formal, periodic reassessments of the plan.

Although the use of the Risk Management Plan should become institutionalized on a project, development of the plan can vary greatly. For smaller projects, a few hours may only be required to conduct a planning session and develop a plan. This time proportionately rises as projects get bigger and more complex. Tens of hours may be necessary to devise a quality Risk Management Plan for a team in charge of a large and complex project.

Issues and Risks

What is the difference between issues and risks? These terms are commonly used interchangeably. Some project managers believe that risks and issues define different concerns and should, therefore, be defined into different categories that need different managerial responses.

An issue is an event that has already happened. For example, a loss of a team member is an issue that led to a month delay. In contrast, risk can be characterized as what could happen to the detriment of the project. For example, "a possibility of losing the project manager could cause a late completion of the project." Risks are in the future. Consequently, while we strive to *resolve* an issue, our managerial response is to *prevent* a risk or its impact.

For questions such as: "What are the issues causing the variance?," the aim is to identify what has happened that caused the schedule variance and needs to be treated. On the other hand, answers to a question such as: "What new risks may pop up in the future and how could they change the preliminary predicted completion date?" seek to find future candidate events that need to be acted upon in order to mitigate their impact on the project.

RISK IDENTIFICATION CHECKLIST

Once the Risk Management Plan is developed, the first step in the risk management process is to identify all of the events that could possibly affect the success of the project. Although risk identification is the first step in the risk management process, it is not a one-time event. Risk identification is an iterative process that occurs throughout the project life cycle. Some project teams first begin by identifying the categories of risk, such as technology risk, market risk, business risk, and human risk, and then use brainstorming and other problem identifying techniques to identify all potential risk events within each category.

The key element of this step is to attempt to identify *all* potential risks. Do not make judgment at this step on whether a risk is of real concern or not, that is the next step in the process. When risk identification is done well, it can be overwhelming, especially early in a project when the number of uncertainties is at the highest. Remember that the goal of risk identification is to flush out as many potential risks as possible in order to get them on the table for discussion.

The Risk Identification Checklist is a very good tool to use as a guide and framework for identifying different categories of risk, such as risk coming from the business environment, resource and collaboration activities, and stakeholders to name a few.

Developing a Risk Identification Checklist

The Risk Identification Checklist will be unique for every organization because every organization has a unique set of uncertainties associated with the business environment it operates within, the policies and practices which guide it operation, constraints that affect its project teams, and its ability to access and use information needed to inform the team about future events. Table 10.5 contains a sample Risk Identification Checklist that can be used as a starting point and reference for developing your own checklist.

Table 10.5: Sample Risk Identification Checklist	
Project Management Risks	
☑ Schedule activities are overly optimistic	☑ Timeline assumes the use of specific resources who may not be available
☑ Effort is greater than estimated	☑ Target end date has moved up with no adjustment to scope, time, or cost
☑ Requirements have not been baselined and continue to change	☑ Budget is not based on structured estimates
☑ Functional requirements lack user involvement and input	☑ Risk response plans have not been developed

(Continued)

Table 10.5: (Continued)

Project Management Risks

☑ Lack of performance measures and/or performance reporting process	☑ Project scope, vision, and objectives are not clearly defined
☑ Project does not have senior management or customer buy-in	☑ Other similar projects have been delayed or canceled
☑ Person-hours (hours per month) are not reasonable for the work estimated	☑ All dependencies between functional groups have not been identified

Resource Risks

☑ Hiring is taking longer than expected	☑ The personnel most qualified to work the project are not available
☑ There is tension between the project team and the client	☑ Unexpected training is needed to build required skill
☑ Estimated staffing profile does not seem reasonable given project scope and/or complexity	☑ No resource ramp time was included in the project schedule

Stakeholder Risks

☑ End user rejects project outcome, resulting in rework	☑ End user input is not solicited
☑ End user or client will not participate in review cycles	☑ Communication time between project team and end users or client is slower than expected

Technical Risks

☑ Necessary functionality cannot be implemented using selected technologies	☑ Components developed separately cannot be easily integrated
☑ Quality assurance activities are being ignored or minimized	☑ Inaccurate quality checking may result in quality problems
☑ The programming languages or other technologies are unfamiliar	☑ Development tools are not in place or not working as expected
☑ There is a dependency on a technology that is still under development	☑ The technology being used or developed is new to the organization

Environmental Risks

☑ The project depends on government regulations	☑ Project depends on industry standards which may change
☑ Project deliverables are developed by third parties (subcontractors)	☑ Project interdependencies with external parties exist

Developing a standard set of risk identification categories and items is a good practice as it drives consistency in identifying various risks which can affect project outcomes. The questions contained in the checklist can be developed by first understanding the various work activities contained in the WBS, the constraints within which the project will have to operate, and information contained in other guiding project artifacts such as the Project Business Case (Chapter 2) and Project Charter (Chapter 3). Additional questions can then be developed by tapping into historical learnings through the review of risk events and issues that affected previous projects.

Using a Risk Identification Checklist

Effective risk management begins with thorough risk identification. The ability to deal with uncertainties ahead of time is contingent upon a project team's ability to predict their potential occurrence. That ability is rooted in risk identification.

Ideally, the Risk Identification Checklist is a standard job aid created by members of the project office. If this is not the case, it should be developed early in the project. The checklist is used as a guide for all project participants who will be involved in predicting possible risk events. It is meant to assist the project manager in ensuring various perspectives of risk are considered such as risks associated with managing the project, environmental risks, resource and collaboration risks, risks associated with stakeholders, and certain technological risks when applicable.

The checklist should be distributed to all members of the project team, as well as key stakeholders, who are charged with identifying potential risks for the project. Best practice usage involves continued use of the checklist as a project progresses through its natural project cycle as new uncertainties arise.

To become institutionalized within an organization, the Risk Identification Checklist should be periodically updated to reflect common risk events encountered across an organization's projects. An opportunistic time to update the checklist is during the Postmortem Review process for each project (Chapter 17). As the issues and risks encountered for a project are reviewed, those issues and risks which have a systemic propensity to affect future projects can be added to the Risk Identification Checklist for future reference and use.

RISK REGISTER

The Risk Register provides a record of identified risks relating to a project and serves as the central repository for all open and closed risk events. The risk register typically includes a description of each risk event, a risk event identifier, risk assessment outcome, a description of the planned response, and summary of actions taken, and current status. Many times risk events are prioritized in the Risk Register based upon the risk assessment score or qualitative analysis.

The register provides a project manager and the organization as a whole a central risk knowledge repository. Not only can the knowledge about project uncertainty contained in the Risk Register be used to manage risks on a particular project, it can also be used to mitigate systemic risks which plague many projects within an organization.

Creating a Risk Register

The Risk Register is arguably the most crucial tool for managing project risk. A good register contains all the necessary information about the project risks, provides a comprehensive catalog of the risks, provides a severity determination, and describes the possible responses to the risk events.

The information in the Risk Register can be represented in a number of ways, such as a database, a paragraph-style document, or a spreadsheet. The spreadsheet style, as shown in Table 10.6, is the most commonly used format because it presents all the information pertaining to project risks without the user having to scroll through several pages. The following elements should be included in a Risk Register.

Risk Identifier

Each risk event should have a unique identifier for cataloging and monitoring purposes. The most common approach is to assign each risk a chronological number. Another approach is to map risk events to the WBS element for which they are associated. For instance, risks associated with a level 3 WBS element might have identification numbers of 3.0.1.1, 3.0.1.2, 3.0.1.3, and so on.

Risk Description

The main component of the Risk Register is the risk description. We recommend using an "IF/THEN" format for your risk descriptions (see Table 10.6). The format not only describes the risk, but also describes the potential consequences: "IF" *this* occurs (risk event), "THEN" that will be the outcome (consequences). Another option is "BECAUSE" of (event or condition), (risk event) could occur, "CAUSING" (consequence).

Dates

For risk timing, aging, and tracking purposes, the Risk Register must have a date component. The most common and useful dates are the date on which the risk was identified, the risk trigger date (when the risk is likely to occur), and the closure date.

Severity

In order to prioritize the risk events (remember, you can't address every risk event identified), a severity component should be included in the Risk Register. Either quantitative (1, 2, 3) or qualitative (High, Medium, Low) representation of risk severity is an acceptable approach. Specific definitions for the numerical values or qualitative values are documented in the Risk Management Plan.

Remember to evaluate the severity of risk from two perspectives: (1) the probability a risk event will occur, and (2) the severity of the impact if it does indeed occur. Total risk severity must factor in both probability and impact perspectives.

Response

For each risk event a project team decides to manage, a response approach must be decided upon and documented in the register for reference and tracking purposes. For

Table 10.6: Example Project Risk Register

Risk ID	Risk Description		Dates				Analysis				Response & Action	Owner	Status
	If	Then	Opened	Trigger	Closed		Probability	Impact	Severity				
1	User experience designers are not released from their current project in two weeks. . .	The project kick-off will be delayed two weeks	3/12	3/28	3/22		5	2	10 (HIGH)	Avoid: Request release of resources at next portfolio approval meeting	Ranger	Closed. Portfolio decision body approved the hiring to three additional people
2	Insufficient digital data storage capacity is available for weekend customer transactions. . .	The system will experience unscheduled down time of up to 60 hours	2/28	4/22			3	4	12 (HIGH)	Mitigate: Enable system transaction limits over each weekend until additional storage is available	Jordan	Active: Transaction limit feature in development. Request for quotes for additional storage have been released
3	Primary stakeholders to not agree on proposed product price. . .	Features will need to be removed from the design	1/29	5/1			2	2	4 (LOW)	Accept:	Harkin	Inactive: Risk deemed as low risk. Will continue to monitor on a monthly basis

low-priority risks and others that the team decides not to manage, the default response is *acceptance*. The Risk Register must contain a field to identify the chosen response for each risk event.

Owner

Every risk event, regardless of priority, must have an owner assigned. The Risk Register therefore must provide an owner component. The risk owner is the person who is responsible for monitoring the risk event and initiating the risk response action if and when it is necessary.

Status

Risk events are dynamic by nature, meaning they can change state over time. To facilitate communication, a Risk Register should include a risk status field. The most common risk statuses include open, monitoring trigger event, response initiated, and closed.

The Risk Register is a very flexible tool in that it can be constructed with any number of components as stated previously. Upon initiation of a project, take time to design the Risk Register format and components that support the risk management methodology described in the Risk Management Plan.

Using the Risk Register

As the central tool for managing risk on a project, the Risk Register has many valuable usages. First, the register serves as the central repository for all risk events. Since it catalogs all project risks, its use must be started in the earliest stage of a project and used throughout the project life cycle. Identifying new risks and updating the Risk Register should be part of an ongoing risk management process.

Since all risk events for a project are contained with the register, the opportunity exists to use the tool to prioritize the risk events. Since most projects contain more risk than a project has resources to manage, trade-off decisions have to be made concerning which risk events to manage and which to either accept or simply monitor. The Risk Register provides the necessary data and structure to represent risk event priority. Normal practice is to put focus on the top three to ten risks at any one time.

The Risk Register also fosters risk-related communication with project stakeholders. This can either be accomplished by using the register itself as a communication device, or by using select information within the register to feed other project communication tools. Used in its entirety, the Risk Register can be used to communicate the overall risk profile for a project.

Since most risk events have consequences, the Risk Register is also an effective tool for assisting project managers in developing budget and schedule risk reserve. By effectively using the IF/THEN approach discussed earlier to describe risk events, evaluation of the potential exposure of the highest priority risks provides a minimum, most likely, and maximum range for exposure for the project.

Lastly, the Risk Register is used to periodically monitor the status of risk events which have been identified. Project team members are balancing many tasks at any given time, so they need an ongoing process and tool to remind them of their risk management duties. With effective use of dates within the register, the project manager is able to track the overall risk trend of a project.

Risk trend charts track the overall project risk profile over time. If risk is being managed effectively, the project risk severity should decrease over time as illustrated in the figure. If not, additional resources may be needed to specifically resolve risk events, or an evaluation concerning project terminations should be considered.

RISK ASSESSMENT

When you have identified risks that could affect your project outcomes, you need to determine which ones you will spend project budget, time, and resource effort managing as not all risk events require action. The risk assessment step is needed to sift through all of the risk events identified and determine those that pose the most serious threat to the success of the project. The result is a prioritized "short" list of project risks that the team can then manage. The Risk Assessment Matrix is useful for increasing visibility and awareness on the part of top managers within an organization of the critical risks associated with a project. This enables sound decisions pertaining to the risk to be made in the appropriate context.

We recommend that a project team begin with a qualitative approach to risk assessment, at least for the first iteration of analysis. By this we mean assessing whether the severity of impact and the probability of occurrence is high, medium, or low for each of the risks identified. This gross analysis will accomplish two important things. First it will quickly prioritize the risks, so the highest risk can be identified for immediate action. Second, it gives the project manager and team an understanding of the overall risk level of the project. The Probability and Impact Matrix is an excellent tool for this type of risk assessment (see Table 10.4).

We described how to develop a generic Probability and Impact Matrix when discussing the Risk Management Plan. Therefore, we will look at tailoring risk assessment to meet the needs of the project.

Tailoring Risk Assessment

One way to tailor the Probability and Impact Matrix to meet your needs is to adjust the probability and impact scales. The scale for impact is shown using a rating from 1 to 5 in Table 10.2. You can modify these to emphasize one project objective differently than the others.

Suppose you are working on a critical project with a very tight timeline. There are two ways you can modify the impact scales to reflect this. The first is to modify the schedule impact parameters as shown in Table 10.7. You can see with the modified schedule impact there is very little room for schedule delay.

Another option is to modify the impact scale. Instead of using a rating scale of 1–5, you can have each step up double the impact, as shown in Table 10.8. This method emphasizes the impact of schedule delay.

Other options are to use a scale of 1, 3, 5, 7, 9, or use three ratings of 1, 3, 9. The 1, 3, 9 option immediately draws attention to significant schedule delays.

Understanding the probability and impact of a risk is a good first step in getting a good understanding of a risk event, but you don't have to stop there. There are many other variables you can consider to assist in prioritizing risks. These three variables are helpful in getting a fuller understanding of a risk.

- *Proximity*. Proximity refers to when the risk event could occur. Risks that have a high proximity are likely to happen sooner than those with a low proximity. Risks with a high proximity should receive higher priority than those with lower proximity. Proximity may be referred to as the impact window.
- *Urgency*. Urgency references the window of time you have to implement a response. Higher urgency indicates you have less time to implement a response in order for it to be effective. These risk responses should be prioritized.
- *Manageability*. Manageability indicates how easy it is to address the risk. Those risks that have low manageability should be prioritized as it may take significant time and resources to implement an effective response.

Table 10.7: Modified Schedule Impact Parameters

Scale	1 Very Low	2 Low	3 Medium	4 High	5 Very High
Generic Schedule Impact	Slight schedule delay	Overall project delay <5 %	Overall project delay 5–14%	Overall project delay 15–25%	Overall project delay >25%
Modified Schedule Impact	No delay	Overall project delay of 3 days	Overall project delay of 1 week	Overall project delay of 2 weeks	Overall project delay of > 2 weeks

Table 10.8: Modified Impact Scale

Scale	.5 Very Low	1 Low	2 Medium	4 High	8 Very High
Schedule impact	Slight schedule delay	Overall project delay <5%	Overall project delay 5–14%	Overall project delay 15–25%	Overall project delay >25%

Using the Risk Assessment

If risk identification is adequately performed, it is common for a large number of risks to be identified. The challenge is to identify those that should be addressed and those that can be monitored.

The matrix helps identify the highest-ranked risks. For example, focus on handling risks with a risk score of greater than 11 (Table 10.4), and treat other risks as noncritical. With this approach, you optimize your resources to focus on the most significant risks. Noncritical risks are still important, but they may not need to be addressed immediately given scarce project resources.

A Risk Map can be effective for visualizing risk events. Shown in Figure 10.2, the map displays the most critical risks from the two dimensions of likelihood of occurrence and risk impact. We show numbered scales, but others may prefer to choose more qualitative values such as "high," "medium," and "low."

Normally, a threshold line is drawn to visually indicate the separation of the critical risks from those at a lower severity level. The line is determined from the description of the scale levels discussed earlier. It is important to include risk events which are below, but near the threshold, as these are the risks that are most likely to move to a critical state.

The final piece of project intelligence that can be included on a Risk Map is change in likelihood of occurrence or impact. This is indicated by arrows on Figure 10.2.

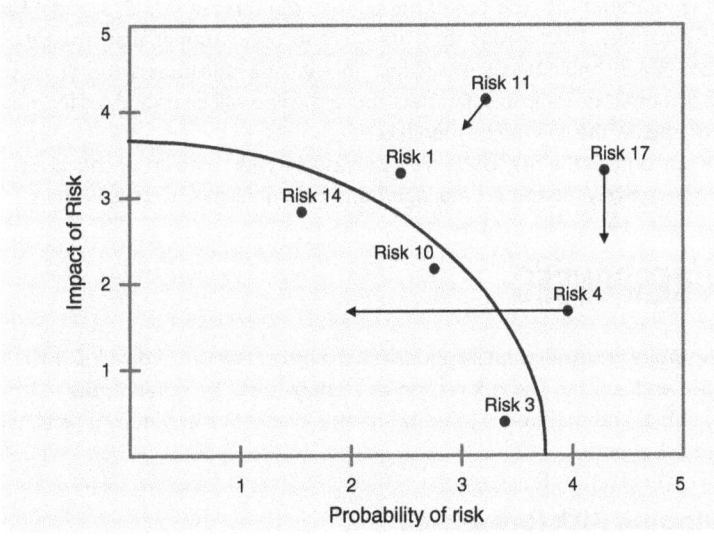

Figure 10.2: Example Risk Map

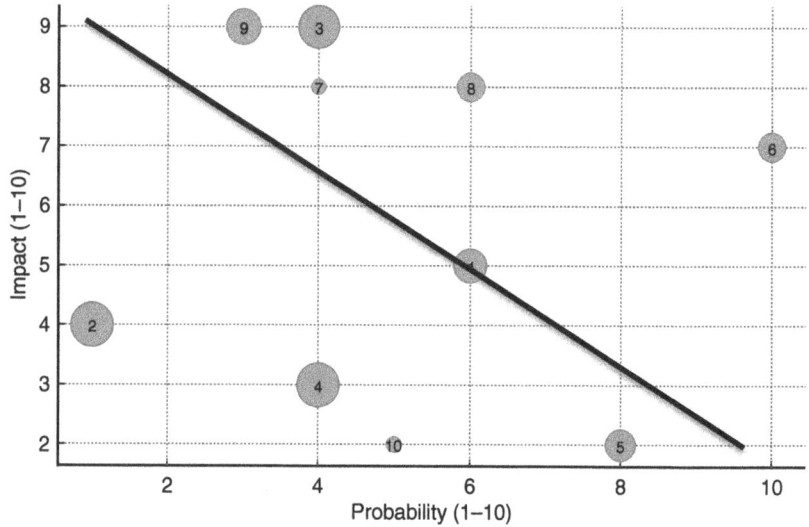

Figure 10.3: Example Bubble Chart

If you are using three variables to assess risks, such as probability, impact, and proximity, you may use a bubble chart, such as the one shown in Figure 10.3.

This chart shows the probability on the *X*-axis, the impact on the *Y*-axis, and the proximity indicated with the size of the bubble. You can see risks 2, 3, and 4 have the highest proximity. That means they are likely to occur sooner than the other risks. Therefore, Risk 3 should be addressed first since it has a high impact and a high proximity. The diagonal line indicates that risks above the line will be prioritized for responses, and those below the line will be monitored.

The outcome of the risk assessment process is a shortened list of critical risk events that can be actively monitored, managed, and communicated to stakeholders.

RISK RESPONSES

With the project risks adequately prioritized the project team can develop risk response strategies and actions. The risk responses strategies will be noted in the risk register, implemented, and monitored to ensure they are effective in reducing or eliminating selected risk events.

Developing Risk Responses

Any suitable risk response action essentially falls into one of the five broad categories of response strategies: avoidance, transference, mitigation, escalation, and acceptance of risk.6

Risk Response Definitions

Risk avoidance. Taking action to eliminate the threat.
Risk mitigation. Reducing the probability and/or impact of a threat.
Risk transference. A risk that will have a positive impact on a project.
Risk escalation. Bringing the risk to someone with more authority to address.
Risk acceptance. Acknowledging the existence of a risk, but taking no action unless the risk occurs.
Secondary risk. A risk that occurs as the result of a risk response.
Risk trigger. An event or condition that indicates that a risk has or is about to occur.
Fallback plan. A risk response used when other responses have failed.

When avoiding a risk, you're taking actions that eliminate the threat. Examples include:

- Eliminate uncertainty. If you have uncertainty associated with a deliverable, you can do more research to eliminate the uncertainty.
- Relax or eliminate a constraint. If you have multiple schedule risks, you can extend the schedule to avoid the schedule constraint that is putting the delivery date at risk.
- Change a deliverable. For risks associated with technology, you might be able to find a proven or less-risky technology.

For all risk responses, but especially when avoiding risks, we sometimes introduce a new risk based on the response. This is called a secondary risk.

Mitigating a risk means finding ways to reduce the probability of an event, and/or the impact of the event if it does occur. Options for mitigating threats include:

- Developing prototypes or models of products
- Running additional tests
- Having redundant systems.

Transferring risk usually equates to spending money to shift the risk management to another party. The two most common means of transference are insurance and contracts.

- *Insurance*. With an insurance policy you pay an insurer to absorb the financial risk of an uncertain event. However, be aware that if the event occurs, it is still your problem, the insurance company merely pays the bill.
- *Contracts*. You can transfer the work and the management of the risk to a vendor who might be better qualified to manage the risk. However, if the vendor isn't effective, you are still accountable for the result.

Risk transference is another example where secondary risks associated with vendors can occur.

Escalation is used when the risk event is outside the scope of the project, or when you don't have the authority to implement the response. Examples of when escalation is an appropriate response include:

- When the project is part of a program, and the threat could impact several projects in the program.
- If the response could impact departments or divisions that are not involved with the project.
- If the cost of the threat response exceeds the project manager's budget authority.

Threats are usually escalated to the sponsor, PMO, program manager, or product owner.

For those risks that are not among the highest-ranked risks, or for risks that have no other viable response strategy, a risk acceptance strategy is used. This implies that project managers have decided to not change the project plan or are unable to articulate a feasible response action to deal with a risk.

Risk acceptance can be passive or active. With passive acceptance if the event occurs, you deal with it in the moment. A common acceptance strategy is to have reserve time in the schedule or reserve funds in the budget to absorb overruns.

You can also take a more active approach to acceptance, such as developing a contingency plan to implement in the event the risk materializes. When you have a contingency plan, it is useful to identify a risk trigger that lets you know that the risk is imminent.

In some cases, you may want to develop a fallback plan in case other risk responses are not effective. For example, if you are upgrading a system, and it is causing a lot of problems, you can fall back on the old system.

Contingency Reserve

Contingency reserve, also known as risk reserve, is the funds or time incorporated into a baseline to accommodate known risks and risk responses. There are different ways to estimate the amount needed for reserve. The simplest method is to estimate the reserve based on past experience.

Another method is to determine the exposure for each risk by multiplying the probability times the impact, and then sum those numbers to see the risk exposure for the project. Then take a percentage of that number, assuming that some percent of the risks will occur, and use it for reserve.

If using a Monte Carlo simulation (described in the next Chapter), you can use the simulation to indicate the amount of cost or schedule reserve needed to meet a specific target.

For reserves for the unknown unknowns, some firms develop *management reserves* involving cost, schedule, or both to allow for such future situations when cost or schedule objectives may be missed. Once the reserves are used, the cost baseline gets changed. Managing management reserves is in the domain of top management, typically the project sponsor.

It is important to note that you are not limited to a single risk response. You can and should develop multiple responses to reduce the risk of an event to an acceptable level.

Risk responses with an assigned risk owner and activities and applicable budget ensure that the right people are aware of and ready to act should a risk event occur. Assigning a risk owner ensures you have a well-qualified person monitoring the event. This includes watching for and responding to risk triggers.

Implementing Risk Responses

Identifying the risk responses is followed by identifying risk owners—individuals or parties responsible for implementing each response, identifying triggers, and monitoring the effectiveness of the responses. At this point you will likely need to update your schedule and budget to account for risk responses. In some cases, you will also update your requirements and WBS as well.

Some risks are independent, leaving their owners fully responsible for their management, other risks might be interdependent. If so, their actions should be developed and owned interdependently. Once risk responses are implemented you should see a change in the risk assessment. The number of risks with a high score should be fewer and the number with a lower score should increase. The Risk Map in Figure 10.2 shows how risks have moved on the probability and impact matrix based on the implementing responses.

Risk Monitoring

Most of a project manager's attention with respect to risk management tends to focus on the activities associated with risk identification, risk assessment, and risk response planning. Where project managers historically spend less time and focus are the activities associated with risk monitoring. It is not uncommon for project managers to be surprised when a risk event they had identified earlier, but were not monitoring, suddenly occurs. To protect against this, diligent risk monitoring is required. The risk owner should be monitoring those risks assigned to them, and the project manager should be checking with the risk owners on a regular basis to track open and active risks.

The first key aspect of risk monitoring is systematically tracking the status of risks previously identified. You should note if a trigger has been activated, indicating that the risk event has occurred. In this situation responses that have not yet been implemented are activated. You will also want to determine if the planned responses are effective. If not, you may need to develop additional responses. If the risk event has passed, note that the risk is closed on the risk register.

Another key aspect of risk monitoring is identifying, documenting, assessing, and responding to any new risks that emerge. These are called emergent risks. An emergent risk goes through all the steps in the risk management process.

You should monitor the schedule and budget reserve to ensure you have sufficient reserve left to address the remaining risk exposure. When you use reserve to address a threat you should reduce the amount of reserve that is still available, so you always have a clear idea of how much reserve remains.

When a risk event occurs, when responses were effective, or not effective, it is useful to record that information in a Lessons Learned Register. This information will be compiled at the end of the project for use in future projects.

The key to risk management lies in thorough identification and robust response planning, implementation, and monitoring. Planning and assessment without effective response planning and implementation is hardly better than not conducting risk management. Well-thought-out responses help avoid unpleasant outcomes. Keeping the risk register up to date with existing and emerging risks increases the likelihood of delivering on time and within budget.

CHOOSING RISK MANAGEMENT TOOLS

The tools presented in this chapter are designed for various project risk management situations. Matching the tools to their most appropriate usage is sometimes a bit confusing. To help in this effort, the following table lists various risk management situations and identifies which tools are geared for each situation.

Situation	Risk Mgmt Plan	Risk ID Checklist	Risk Register	Risk Assessment	Risk Responses
Establishes risk management methodology for a project	✓				
Supports the identification of project risks		✓	✓		
Establishes risk categories		✓			
Establishes a project risk repository			✓		
Used to determine risk severity and prioritization				✓	
Provides focus on the most critical risks			✓	✓	✓
Provides information on risk timing			✓	✓	
Supports qualitative risk analysis				✓	
Used to help determine if action and resources are required				✓	✓
Documents risk response options			✓		✓
Used to determine risk reserve				✓	✓
Monitors risk status			✓		✓

References

1. Gilb, T. (2005). *Competitive Engineering: A Handbook for Systems Engineering, Requirements Engineering and Software Engineering.* Oxford, UK: Butterworth-Heinmann.

2. Martinelli, R., Waddell, J., and Rahschulte, T. (2014). *Program Management for Improved Business Results,* 2nd ed. Hoboken, NJ: John Wiley and Sons.

3. Snyder Dionisio, C. (2023). *Hybrid Project Management.* Hoboken, NJ: John Wiley and Sons.

4. Sunstein, C.R. and Mastie, R. (2014). *Wiser: Getting Beyond Group Think to Make Groups Smarter.* Boston, MA: Harvard Business Review Press.

5. Duckert, G.H. (2010). *Practicing Enterprise Risk Management: A Business Process Approach.* Hoboken, NJ: John Wiley and Sons.

6. Snyder Dionisio, C. (2023). *Hybrid Project Management.* Hoboken, NJ: John Wiley and Sons.

PART

IV

Project Implementation Tools

11

ADVANCED RISK MANAGEMENT TOOLS

All projects should engage in risk management planning, risk identification, qualitative risk assessment, and planning risk responses as discussed in Chapter 10. Large, complex, and "risky" projects can benefit from quantitative risk assessment. These types of projects are also more likely to utilize risk reports and risk dashboards to convey information about the risk status of the project. This chapter will focus on the advanced risk management tools that can be used on these projects.

There are two aspects of quantitative risk assessment. One aspect involves analyzing individual risk events numerically. Similar to qualitative risk assessment that was covered in the previous chapter, quantitative risk assessment assesses the probability and impact. However, it requires more knowledge and precision than qualitative assessment. When conducting a quantitative assessment you generally use historical data to determine the probability of an event occurring. For example, if a risk event has occurred 6 out of the last 10 times on a similar project, the probability is 60%. The impacts are focused on time and budget. Therefore, you are assessing the time delay or the budget overrun.

Another aspect of quantitative risk assessment is evaluating the likelihood of meeting the cost and schedule objectives for the project. This involves looking at all the risk events and other sources of uncertainty to determine the probability of achieving the target delivery data and the estimated budget. Other sources of uncertainty can include volatility in material cost or availability, ambiguity associated with future events, complexity due to the number of systems and stakeholders involved, and general uncertainty. An honest assessment of the uncertainty facing the project can help you determine the appropriate amount of cost and schedule reserve needed to establish a realistic schedule and budget.

There are several ways to assess risk quantitatively. We are going to utilize decision trees and a Monte Carlo analysis.

Project Management Toolbox: Tools and Techniques for the Practicing Project Manager,
Third Edition. Cynthia Snyder Dionisio, and Russ J. Martinelli.
© 2025 John Wiley & Sons, Inc. Published 2025 by John Wiley & Sons, Inc.

MONTE CARLO ANALYSIS

While managing a project, a project manager will face the situation where they have a long list of risk events, and little clue of the impact they may have on the project goals. Once the risk events have been assessed, a Monte Carlo Analysis can be performed to quantifiably evaluate the potential impact of all the risks on the project cost and schedule objectives.

Monte Carlo Analysis is effective in determining the impact of identified risks by running mathematical simulations to identify a range of outcomes associated with various confidence levels relating to probability of success. The simulation furnishes the project manager with a range of possible outcomes and the probability they will occur for any choice of action. This process provides a valuable tool to compensate for the impact of critical risk events by determining the amount of risk reserve needed to increase the probability of success given the known uncertainties facing a project.

A Monte Carlo Analysis randomly samples a probability distribution of the critical project risks to simulate project scenarios in response to the risk impacts hundreds or even thousands of times. This provides statistical distribution of the calculated project durations and approximates the expected value of the duration, as illustrated in Figure 11.1. With these distributions, you can quantify the risk of various schedule scenarios, alternative implementation strategies, activity paths, or even individual activities. For example, as Figure 11.1 indicates, there is a 40% probability that the project will be finished before or on May 20.[1]

Performing a Monte Carlo Analysis

A Monte Carlo analysis generates graphical representation of multiple outcomes. The graphical analysis enhances evaluation and communication of decision choices. Typically, Monte Carlo Analysis deals with schedule and cost risks, although other facets such as the quality of the final project output can at times be analyzed. Taken overall, performing a schedule risk analysis is more complex than a cost analysis, simply because dependencies between project activities need to be established in order to identify the

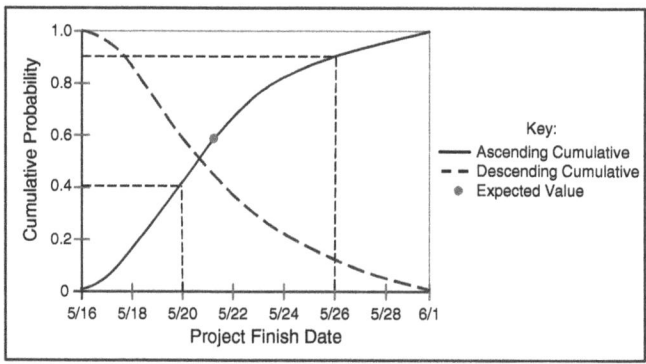

Figure 11.1: Cumulative Distribution of Project Duration Produced in Monte Carlo Analysis

critical path. For this reason, our focus is on looking at the Monte Carlo Analysis process in schedule risk analysis (see Figure 11.2).

A number of important inputs are needed to perform a successful Monte Carlo Analysis. First, the Risk Management Plan should provide guidance on when and how to apply Monte Carlo on a project. The critical project risks and their estimated impact to the project provide the risk impact from which the simulations are built upon. Finally, the project schedule or budget for the project is needed to establish the baseline from which probability impact scenarios will be added. When analyzing impact to project timeline, the Gantt chart schedule format is the most useful (Figure 11.3).

This information will be fed into software that can run a Monte Carlo Analysis to generate a range of possible project durations (see "Basic Terminology of Monte Carlo Analysis").

Basic Terminology of Monte Carlo Analysis

Chance event is a process or measurement for which we do not know the outcome in advance.

Continuous distribution is used to represent any value within a defined range of values (domain).

Discrete distribution may take one of a set of identifiable values, each of which has a calculable probability of occurrence.

Deterministic model is where all parameters are fixed, having single-valued estimates.

Expected value (EV) is the probability weighted average of all possible outcomes. Synonyms: mean, average.

Mode is the particular outcome that is most likely; the highest point on a probability distribution curve.

Model is a simplified representation of a system of interest such as project critical path chart. It projects project outcome (e.g., project duration) and outcome value (e.g., 18 months).

Probability is the likelihood of an event occurring, expressed as a number from 0 to 1 (or equivalent percentages). Synonyms: likelihood, chance, odds.

Probability distribution represents mathematically or graphically the range of values (e.g., from 2 to 14 days) the variable (e.g., activity duration) can take, together with the likelihood that the variable will take any specific value. Synonyms: probability density function, probability function.

Project scenario is a future state of the project. Synonyms: iteration, trial.

Random sampling is a process generating a random number between 0 and 1, which determines the value of the input variable from the probability distribution.

Random variable is a measure of a chance event. Synonyms: chance variable, stochastic variable.

Single-valued estimate has one value only. Synonym: point estimate.

Standard deviation is the square root of the variance.

Stochastic model is a model that includes random variables. Synonym: probabilistic model.

Variance is the expected value of the sum of squared deviations from the mean.

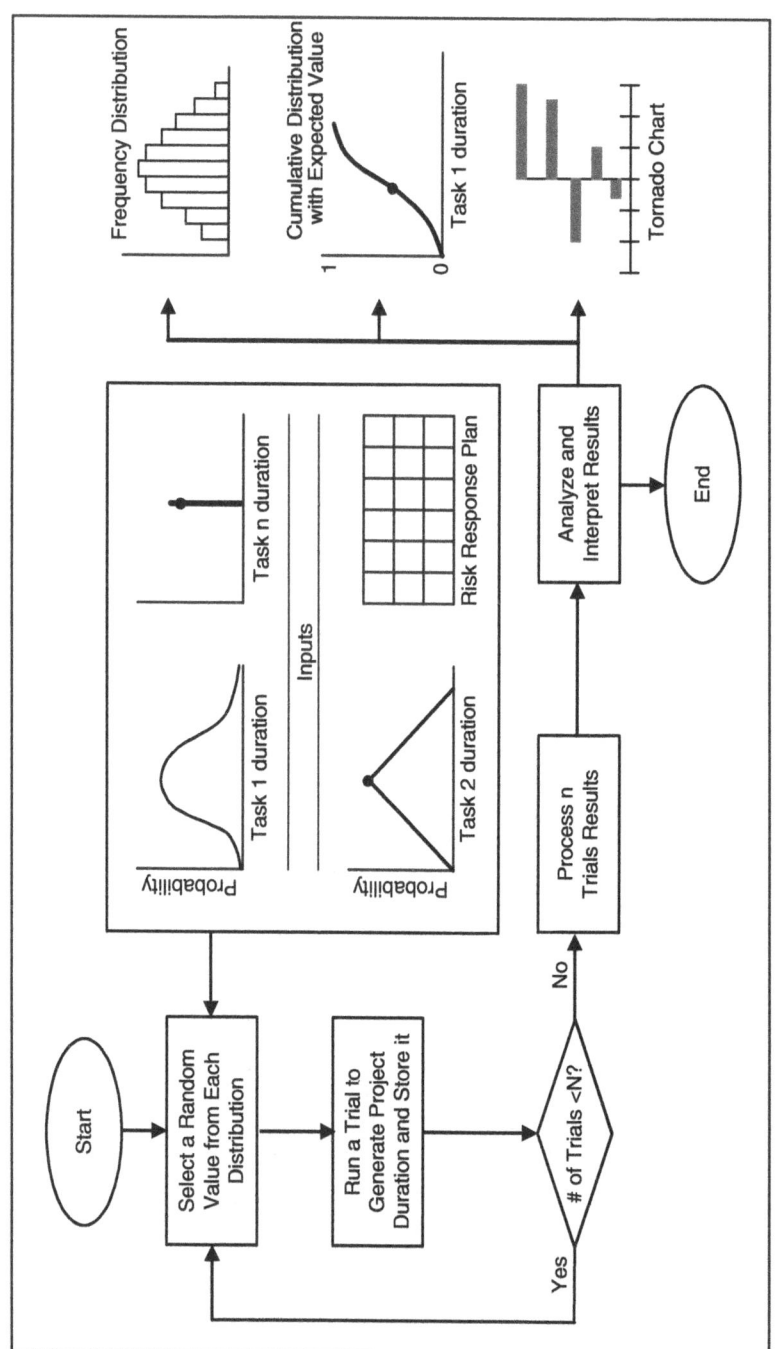

Figure 11.2: Monte Carlo Analysis Process for Schedule Risk

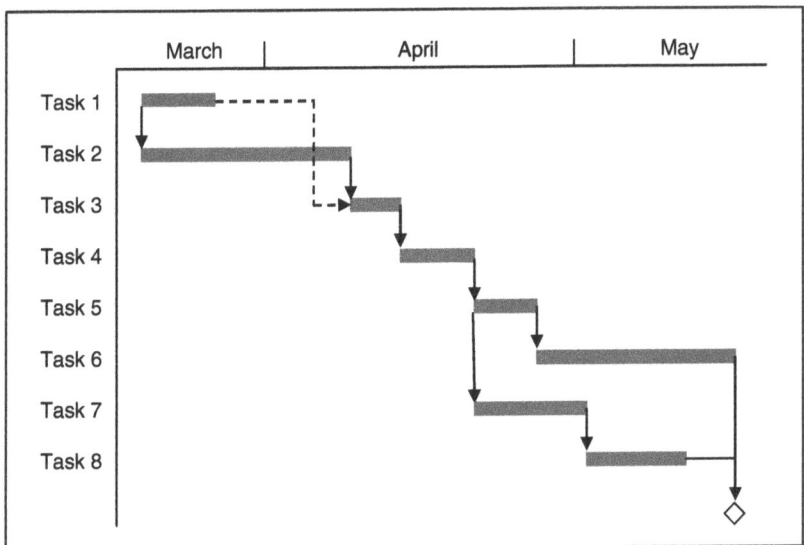

Figure 11.3: An Example of a Gantt Chart for Risk Analysis with Monte Carlo

Generating a range of possible project durations and their probabilities is not possible without preparing probability distributions for project activity durations. This preparation process may begin with the question "How long does it take to complete a project activity?" Let's assume that you performed an activity many times and each time it took ten days to complete. If asked to estimate the duration of that same activity in a future project, you would likely put it at ten days. If each project activity would have such a single point estimate as an input to calculate project schedule duration, the duration would also have only one value. There is not much uncertainty in project activity durations in this single-valued deterministic model—they are all fixed. In the majority of today's projects, such a scenario is not realistic. More realistic is the following probabilistic (stochastic) model.

Imagine that you repeated Activity1 a large number of times (trials, iterations, scenarios), and its duration ran from 5 to 39 days (range of outcomes) due to various issues that arose. You recorded the fraction of times that each duration value (outcome) occurred. The fraction for a particular outcome is approximately equal to its probability (p) of occurrence for Activity 1. When we have these approximate probabilities (the more trials you do, the closer the fraction becomes to the true probability) for all possible outcomes, we can chart them as probability distributions (see the Activity 1 duration curve in Figure 11.2). Assume that experience-based probability distributions are also available for some other activities in the project as well (see Activity 2 duration in Figure 11.2). If we really had such probability distributions, they would be close to objective probabilities, which are defined as being determined from complete knowledge of the system and are not affected by personal beliefs (see "Frequently Used Probability Distributions").

Frequently Used Probability Distributions

Three values are used to describe a very simple and popular triangular distribution (see Figure 11.4a): Triangular (5, 10, 20); minimum (L = 5); most likely M = 10); and maximum (H = 20). Numbers in parentheses are project activity durations in days. The mean is calculated by (L + M + H)/3.

The Beta Distribution (see Figure 11.4b) requires the same three parameters as triangular distribution: minimum (5), most likely (10), and maximum (20). The mean is calculated by (L +4M + H)/6.

Two parameters describe a lognormal distribution (see Figure 11.4c)—mean (10) and standard deviation (2).

Known for its flexibility, the general distribution (see Figure 11.4d) allows shaping the distribution to reflect the opinion of experts. It is described by an array of values (7, 10, 15) with probabilities (2, 3, 1) that fall between the minimum (5) and maximum (20).

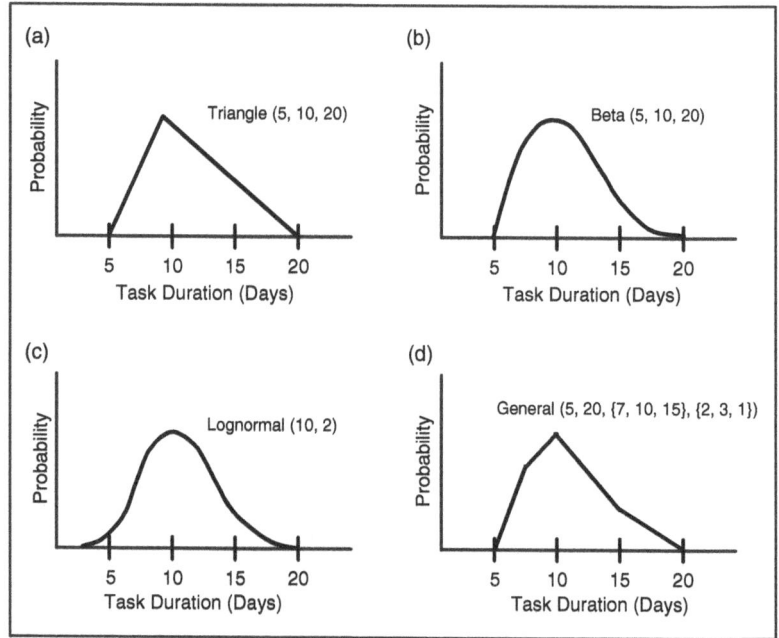

Figure 11.4: Frequently used Distribution. (a) Triangular Distribution, (b) Beta Distribution, (c) Lognormal Distribution, (d) General Distribution

Although some companies do have experience-based databases with the approximate distributions of their project activities, that is the exception rather than the rule. What, then, do we do to prepare probability distributions for activity durations? We will do what is a dominant practice in real-world projects—prepare and rely on subjective probabilities—someone's belief whether an outcome (activity duration) will occur.

The most adequate way for this is to enlist the help of experts, or experienced project participants. Brainstorming with activity owners, studying durations of similar activities in past projects, and consulting other specialists in the company who are not involved in the project all help determine the probability distributions or single values for the activity durations. The single value estimates are incorporated into the baseline project schedule.

Now it is time to apply risk impact. The baseline schedule which was generated from single-value estimates represents the best-case scenario. Monte Carlo Analysis proves a best-case to worst-case distribution of potential outcomes based upon a probability distribution; therefore, we must also define the worst case. This is where the high-level risk events come in to play. Referring to the Probability and Impact Matrix, locate all the HIGH severity risks. The combined schedule impact of the HIGH severity risks represents the worst-case scenario.

Randomly Select a Value from Each Distribution

When the probability distributions are available for the project tasks, select one duration value randomly within the specific range of duration values bound by the best- and worst-case values. The key word here is *randomly*. Using a random sampling technique, Monte Carlo Analysis generates a random number between 0 and 1, which is fed into a mathematical equation that determines the task duration value to be generated for the distribution.[2] All selected values constitute a random sample of values that will be used to generate project durations. The random sampling from the probability distribution is performed in a manner that reproduces the distribution's shape.

Run a Trial to Generate Project Duration

Having a random sample of task duration values means that for each activity in the project schedule there is one value only. Plugging this combination of activity duration values in the project network diagram will produce a scenario for project duration. In essence, this is a deterministic schedule with a single value for project duration, built on single-value durations for each activity. At this time, we will store this project duration until the time comes to use it again.

The software repeats this sequence of random sampling thousands of times. Each iteration produces a plausible outcome. More iterations equate to more confidence in the model. The idea here is that sufficient number of trials preserves the characteristics of the original probability distributions for activities and approximates the solution distributions for project duration.[3]

Process Results

When the trials are complete, our "storage" will contain N project durations. Each one is a possible case for the behavior of the project schedule. Processing them by means of a software program can produce many forms of results. The following are charts are commonly used outputs.

- *Cumulative distribution*. Cumulative distribution, as shown in Figure 11.5, shows the likelihood (shown on the y-axis) of the outcome being less than or equal to a value on the x axis. The dot represents the expected value (or average) given the data.

■ *Frequency distribution.* The frequency distribution is a histogram plot showing relative frequency obtained by grouping the data generated for project durations into a number of bars or classes. Frequency is the number of values in any class. Dividing the frequency by the total number of values will produce an approximate probability that the project duration will lie in that range (see Figure 11.6).

■ *Tornado chart.* This chart shows the extent to which the uncertainty of the individual activities' duration impacts the uncertainty of project schedule duration (see Figure 11.7). Specifically, the bar represents the degree of impact the activity (input variable) has on the project schedule (model's output). Therefore, the longer the bar, the greater the impact that a project activity has on the project duration. Per standard practice, bars are plotted from top down in decreasing degree of impact. When there are both positive and negative impacts, the chart is a bit reminiscent of a tornado, hence the name.

Figure 11.5: Cumulative Distribution

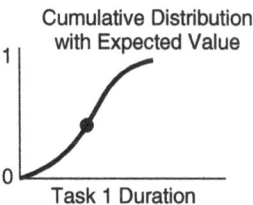

Figure 11.5: Cumulative Distribution

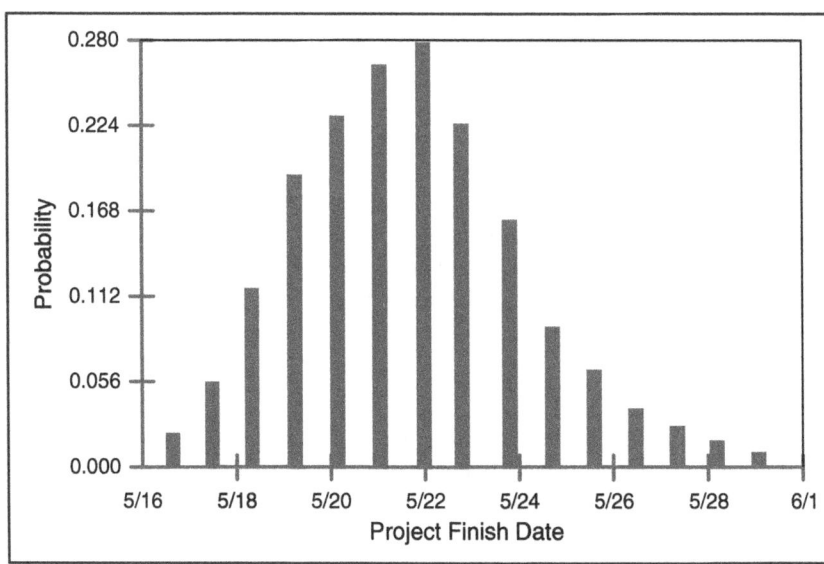

Figure 11.6: Frequency Distribution Histogram of Project Duration

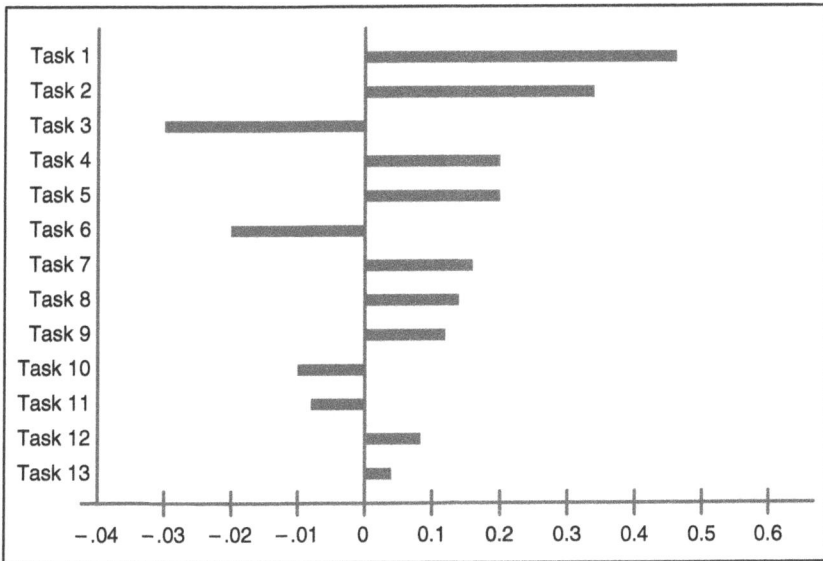

Figure 11.7: An Example Tornado Chart[4]

Analyze and Interpret Results

The results of the schedule risk analysis must be interpreted in a way that clearly provides answers to the questions the analysis was initiated to answer. For that reason, it is beneficial to follow four principles for schedule risk analysis:

1. Focus on the problem
2. Keep statistics to a minimum
3. Use graphs whenever appropriate
4. Understand the model's assumptions.

To demonstrate these principles, let's assume that the project team sets out to perform a schedule risk analysis to answer these questions:

- How likely is it that the team will achieve the project deadline (May 20) imposed by management?
- If the probability is lower than 90% (the team's preferred probability), what do we do to negotiate the deadline with management?
- If we successfully negotiate the deadline issue, which are the top three activities that most impact the project duration?

First, the team goes to the cumulative distribution graph (Figure 11.5). To obtain the answer to the first question, they will:

- Enter the x-axis at the deadline date imposed by management (May 20).
- Move upward to the cumulative curve.
- Move left to the corresponding value on the y-axis. This y-axis value is the probability of completing their project before on the imposed deadline date (40%).

Clearly, the probability is much lower than the preferred 90%. To answer the second question, the team decides to ask management for an option of adding six days of schedule contingency to the imposed deadline, in which case the project would be finished by May 26 and would be 90% probable. To build a better case for negotiations with management, the team developed a strong and clear justification for the contingency by describing the top three risk events that contributed the most potential schedule impact, the probability of success assessment without the contingency (40%), mitigation and avoidance plans to manage the risks, and a call for management action to help the team succeed by increasing the probability of success to 90%.

Finally, the team created a tornado chart (see Figure 11.7) that reveals the three key tasks with the highest impact on the deadline—these are the top priorities to monitor during implementation, an answer to their question three.

Using the Monte Carlo Analysis

Traditionally, it has been the large and complex projects that have most often enjoyed the benefits of Monte Carlo Analysis. For example, if a project is sensitive to a completion deadline, Monte Carlo Analysis is a preferred option. Similarly, if there are many project scenarios and what-if analysis to explore, Monte Carlo is favored over other analysis techniques.[5]

Acquiring the data to run a Monte Carlo Analysis influences how much time it takes to perform an analysis. Interviews with key stakeholders to determine best-case, most likely, and worst-case scenarios can be time consuming. If there is a database with information from previous similar projects, it will take less time.

By taking into account variable factors (such as schedule tasks or cost elements), Monte Carlo makes it possible to spot which variables have the greatest impact on project results. In like manner, the technique makes it possible to analyze the effect of combining multiple variables. For instance, performing project tasks concurrently if additional staff is added versus performing them in sequence.

Monte Carlo Analysis' value is in its ability to examine each project scenario, including the extreme scenarios, to see what conditions give rise to their results. That helps not only to validate the project realism but also to differentiate between what is possible and what is not possible.

DECISION TREE

A Decision Tree is a graphical tool for analyzing project situations that involve uncertainty or risk. It is appropriate when there are only a small number of options or potential outcomes, as opposed to the Monte Carlo analysis that is used for more complex scenarios.

As shown in Figure 11.8, the tree displays sequential decisions in the form of branches of a tree, from left to right, originating from an initial decision point and extending to the end outcomes. The path through the branches represents the sequence of separate decisions and chance events that occur. Decisions are evaluated by calculating the expected value and probabilities of each path. For a description of a typical Decision Tree's components, see "Five Components of a Typical Decision Tree."

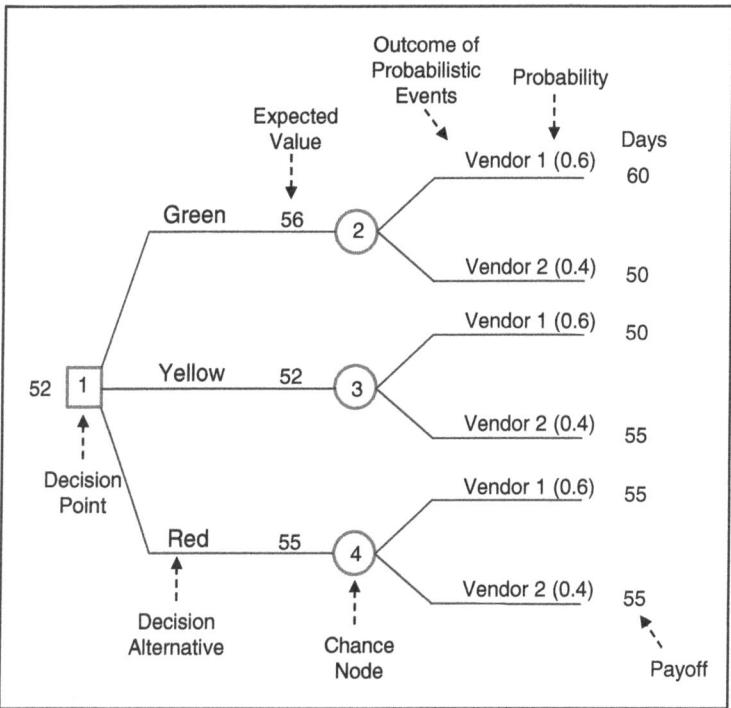

Figure 11.8: Decision Tree for a Project Situation under Risk

Five Components of a Typical Decision Tree

Decision nodes. Also called decision points, these are points in time when decisions are made or alternatives chosen. Shown as square boxes, they are controlled by the decision maker. The starting node is called the *root*.

Chance nodes. These represent times when the result of a probabilistic event occurs. Decision makers have no control over them. Chance nodes are also called probabilistic nodes or points, and are represented by a circle.

(Continued)

Branches. Lines connecting the decision and chance nodes in a sequential manner. The branches leading out of a decision node represent possible decisions, while those stemming from chance nodes represent possible outcomes of probabilistic events.

Probabilities. The probabilities of the probabilistic events shown on the branches representing those events. Mostly they are conditional and for any particular chance node must sum to 1.

Outcome values. Also called the payoff, outcome values of each alternative are placed at the end of the branch. They may represent present values discounted to the date of the root decision or cost.

Building the Decision Tree

There are many different uses for a decision tree which all follow the same process for building the tree. For purposes of this discussion, we will evaluate a decision associated with minimizing schedule duration.

Creating a Decision Tree technique requires input from a number of sources, such as the risk register, duration and cost estimates, the schedule, the budget, information from similar past projects and stakeholder expertise.

Describe the Decision Under Risk

Common sense dictates that in order to make the best decision, we first need to understand the decision context and related risks a project is facing. A convenient way for this is to describe the decision. Here is an example.

Consider a company that competes on time-to-market capabilities. As they are entering the design phase for a new product development project, their goal is to finish it as soon as possible and reach the market before their competitors. While each project day is considered to be of extreme priority, the development cost is given lower importance. In such a situation, the project team is attempting to decide on the appropriate design approach to use for this product. A major uncertainty involves the amount of time it will take to design the central module in the product. Three major alternatives are being considered, each one identified with a single word:

- *Green (G).* Incorporate the routing rules early in the central module design.
- *Yellow (Y).* Predict the routing rules early and modify them at the end of the central module design.
- *Red (R).* Incorporate the routing rules at the end of the central module design.

A second uncertainty is related to an off-the-shelf part that goes into the central module, whichever the alternative. Two different vendors produce the part. It is well known that both companies are in a race to release the newest upgrade of the part, and they have announced the same release date. To represent this decision description and enable its analysis, the team must first structure the model.

Structure the Model

Complex decisions require a level of structure to help simplify the decision options. The Decision Tree model is drawn from left to right beginning with the decision node (the square marked 1). Then the three branches are added to the right-hand side, one for each design alternative—green, yellow, and red (Figure 11.8). Place a chance node (the circles marked 2, 3, and 4) at the end of each branch, followed by two branches, each one for outcomes of probabilistic events—vendor 1 reaching the market first and vendor 2 reaching the market first.

Assess the Probability of the Possible Outcomes

The company's new product development project team cannot wait for the actual release of the first-to-market part by vendor 1 or vendor 2 before beginning their own design process. That would significantly extend the module design schedule, jeopardizing the project end date. Therefore, the project team decides to assess the probability of who—vendor 1 or vendor 2—will release the product first. Their research and past performance of the vendors led them to assess that there is 60% probability that vendor 1's part will reach the market first. The probability for the part from vendor 2 is 40%. These values are added to each of the chance notes. Everybody on the team is clear that these are subjective probabilities influenced by their perceptions, beliefs, and historical events.

Determine the Possible Outcomes

The project team has developed initial network schedules for each of the design options—green, yellow, and red—as if the vendor part is already available. The sequence of design activities involved in each option is different, as well as some other activities. Also, although both vendors' parts can be used for the central module, the process of their incorporation into the design is different, causing the duration of each outcome to differ. Because the team's expectation is that the first-to-market vendor part will be released sometime midway through the module design, they need to evaluate how such a release is going to change the initial network schedules. The product of their evaluation is a set of possible outcomes values, also called payoffs. These schedule durations of the outcomes expressed in days are added at the end of each branch (Figure 11.8).

Evaluate the Alternatives

Once the tree is structured and the outcomes and their respective probabilities are documented, we can evaluate the possible outcomes and select the one with the shortest-possible schedule. To accomplish this, we need to solve the tree.

To solve the tree start at the far right of the tree and work back to the left for each chance node. Calculate the expected value as the sum of each branch's outcome value (payoff) multiplied by the probability.

- Chance Node 2 Expected Value: $(0.60 \times 60 \text{ days}) + (0.40 \times 50 \text{ days}) = 56 \text{ days}$
- Chance Node 3 Expected Value: $(0.60 \times 50 \text{ days}) + (0.40 \times 55 \text{ days}) = 52 \text{ days}$
- Chance Node 4 Expected Value: $(0.60 \times 55 \text{ days}) + (0.40 \times 55 \text{ days}) = 55 \text{ days}$

When the process is completed, the alternative with the best outcome value for the leftmost decision nodes becomes the best alternative. When looking to have the shortest possible schedule, or the lowest possible cost, the best alternative is the one with lowest expected value. In this scenario the shortest schedule is the yellow option with 52 days.

Using Decision Trees

Decision Trees reduce an evaluation and a comparison of all decision alternatives under risk to a single value metric. They can be used for many different scenarios in a project. They can be used during project selection to identify the best investment decision. In this scenario you would determine the net present value for each option, and then look at the areas of uncertainty such as time to market or expected consumer acceptance. For project selection you are looking with the outcome with the highest value.

You can also use a Decision Tree to evaluate options with regards to the lowest possible cost. Similar to the shortest possible duration, you would look at several cost options and the probability of each to calculate the value of each branch. Often a cost decision includes the evaluation of making something in house or outsourcing it, and the various uncertainties associated with each option.

Practicing project managers go to decision trees primarily when they need to swiftly evaluate simple alternatives, pick the best alternative, and go on with their daily routine.

RISK DASHBOARD

Due to the dynamic nature of project risks, risk monitoring must continue throughout the life of a project. New risks will be identified, many anticipated risks will disappear, some risks will be mitigated, and some will change in severity due to a change in probability of occurring or a change in potential impact to a project.

Regularly held risk reviews force consistent risk monitoring and enable repeated risk identification, assessment, analysis, and response planning as a project progresses through the life cycle. To facilitate the risk reviews, a number of tools are available to help project managers efficiently move through the monitoring process. We have discussed several risk tools in this chapter and the previous chapter. The final risk tool we will address is the Risk Dashboard.

Developing the Risk Dashboard

The Risk Dashboard is a business intelligence tool that provides the important risk statistics pertaining to a project. The dashboard helps the project manager and his or her top management assess the health of a project as well as possible issues facing the project. An example Risk Dashboard is presented in Figure 11.9.

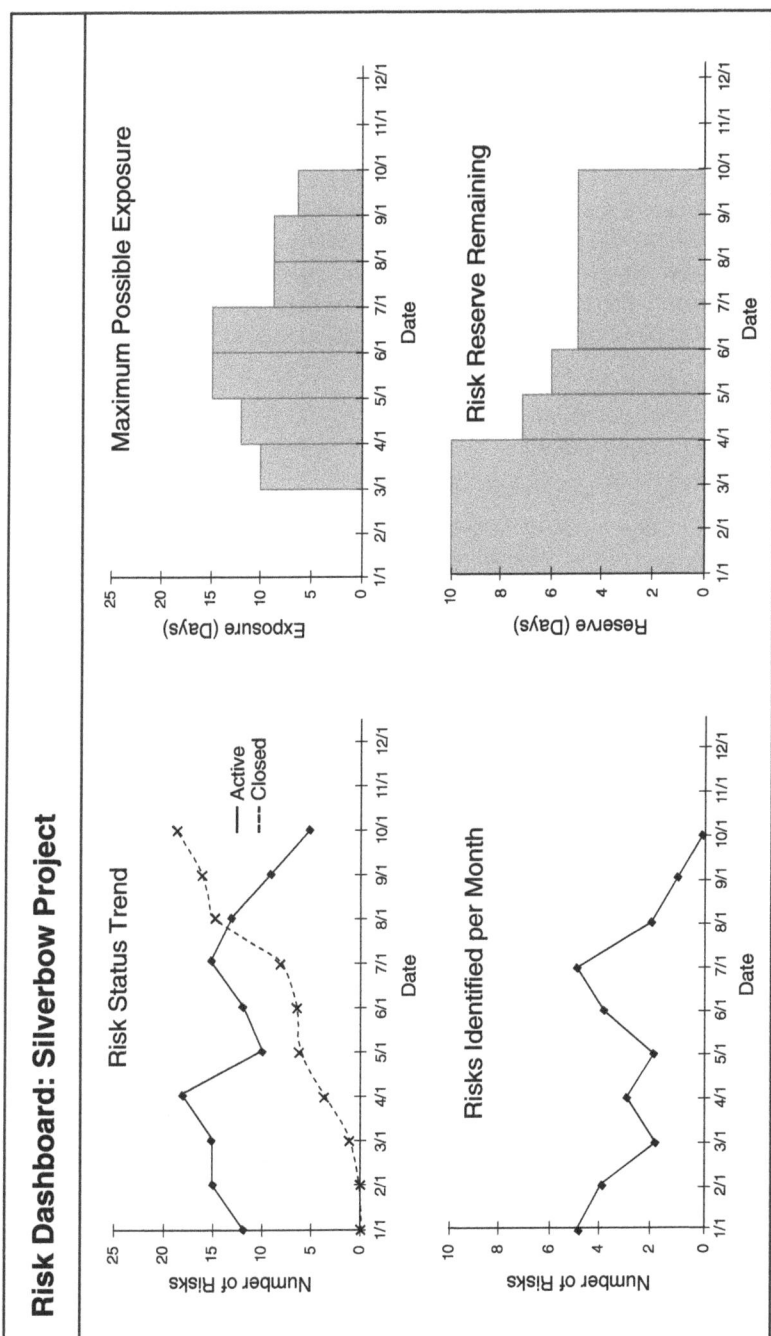

Figure 11.9: Example Risk Dashboard

The Risk Dashboard needs to be carefully designed. Since it is a project and business intelligence tool, you want to ensure that the most important intelligence is being collected, that it is being represented effectively, and that it can be easily and correctly interpreted. All effective dashboards begin with choosing the right set of metrics.

Choosing Risk Metrics

An organization, such as a program or project office, has to develop a risk metric system to effectively measure and monitor the state of risk on each of its projects. Information on the Risk Dashboard is generally provided to senior management periodically and provides top-level information about risk trends and risk status.

As each organization is unique, so too will be the risk metrics it chooses to display. The example Risk Dashboard shown in Figure 11.9 focuses on four primary metrics:

1. Risk status trend
2. Risks identified per month
3. Maximum possible exposure
4. Risk reserve status.

Risk status provides a view of the overall number of risk events open and closed over the life of the project to date. Risks identified per month tells the story of the amount of uncertainty facing the project team over time. The maximum possible exposure indicates the potential impact to the project and business based on the currently identified risk events. Risk reserve status provides information on the amount of risk reserve that has been consumed and what remains at the project manager's disposal.

Create the Dashboard Graphics

With the metrics which are going to be represented in the Risk Dashboard defined, you know what information should be shown in the dashboard. The job of a dashboard is to display a large amount of critical information about a project in a very concise manner, and in a way that decisions can be made.

The Risk Dashboard should therefore provide meaningful information that is appealing to the eye and conveys information simply and quickly. Therefore, it is best to provide only a few key indicators much like the dashboard in an automobile.

Trend information is best represented in a line graph, while comparison of information (such as number of risks that are active, inactive, or closed) is best represented in either bar charts or pie charts.

We recommend using common graph styles and keeping them simple. Remember, the primary consumer of the Risk Dashboard will be your top management team. You will want to spend your time with them talking about the status of risk on your project, not helping them interpret your graphs.

Populate the Dashboard

This final step in development of the dashboard involves collecting, synthesizing, and graphically representing current risk information about your project. While all previous steps are normally performed at the beginning of a project, keeping the information updated is continuous throughout the life of the project.

It is well known that the best risk management practice is that which is performed on a continuous, or at least periodic, basis due to the dynamic nature of project risk. Keeping the information within the Risk Dashboard current is a great way to ensure risk is being actively monitored and managed consistently.

Using the Risk Dashboard

The Risk Dashboard is a tool that effectively provides project and business intelligence. As such, its key usages involve the monitoring of the health of a project, communication of the current risk status and trends, and supporting various risk-based decisions.

A healthy project is one that is on track to meeting its objectives and business goals. That means the project is actively being protected from the potential negative exposure caused by threats. The Risk Dashboard is a good indicator of project health. It supports effective risk monitoring by collecting the critical information about risk status and presenting it in a fashion that conveys project health.

Since the dashboard contains critical information about the project health, it also becomes an important communication device. All project stakeholders have a vested interest in the outcome of a project for various reasons, so the stakeholders also have a vested interest in knowing the health of the project. The Risk Dashboard presents risk status and trend information to project stakeholders in a synthesized and summarized manner. If designed correctly, project stakeholders will be able to determine the health of a project within a few minutes.

CHOOSING ADVANCED RISK MANAGEMENT TOOLS

The tools presented in this chapter are more advanced than the planning tools presented in the previous chapter. They are quantitative tools that are used for specific project risk management situations. To help match the tools to their most appropriate usage, the following table lists various risk management situations and identifies which tools are geared for each situation.

Situation	Monte Carlo	Decision Tree	Risk Dashboard
Provides focus on the most critical risks	✓		✓
Explores project scenarios and what-if analysis	✓	✓	
Used to determine risk reserve	✓		
Monitors risk status			✓
Supports quantitative risk analysis	✓		

References

1. Vose, D. (2012). *Risk Analysis: A Quantitative Guide*, 3rd ed. New York, NY: John Wiley & Sons.
2. Vose, D. (2012). *Risk Analysis: A Quantitative Guide*, 3rd ed. New York, NY: John Wiley & Sons.
3. Schuyler, J. (2018). *Risk and Decision Analysis in Projects*, 3rd ed. Aurora, CO: Planning Press.
4. Vose, D. (2012). *Risk Analysis: A Quantitative Guide*, 3rd ed. New York, NY: John Wiley & Sons.
5. Kendrick, T. (2015). *Identifying and Managing Project Risk*, 4th ed. New York, NY: AMACOM.

12

CHANGE MANAGEMENT

When it comes to managing a project, the one constant you can count on is change. Changes happen, or attempt to happen, at a constant pace on projects. Throughout a project's life cycle, the need and desire for change will come from project team members, the project sponsor, senior leaders, customers and clients, and other project stakeholders.

Each change has the potential to modify the project, whether it is the scope, schedule, cost, resources, or other changes. Most changes impact more than just one objective. Therefore, change management is necessary to determine if each change is necessary, how it will impact the work, and ensuring each necessary change is implemented properly.

Without change management we can get uncontrolled scope change, commonly referred to as scope creep, missed milestones, budget overruns, and unhappy stakeholders.

To help you manage change, this chapter provides high-value tools that should be a part of every project manager's PM Toolbox for establishing and performing project change management activities.

CHANGE MANAGEMENT SYSTEM

Managing change on a project cannot be handled in an ad hoc manner, regardless of the size or complexity of the project. Change management requires the establishment and use of an effective Change Management System.

The Change Management System is to process, document, approve, and manage changes. The Change Management System defines how changes to the project will be managed once a project baseline has been established.[1] Change management protocols will normally include the following:

- Change requests must be made in writing
- The benefits gained from a proposed change must be clearly articulated and documented

Project Management Toolbox: Tools and Techniques for the Practicing Project Manager,
Third Edition. Cynthia Snyder Dionisio, and Russ J. Martinelli.
© 2025 John Wiley & Sons, Inc. Published 2025 by John Wiley & Sons, Inc.

- Change management roles and responsibilities must be established
- An approval process must be documented
- A decision maker has to be appointed
- Approved changes must be incorporated into the applicable project plans and documents
- Changes must be adequately communicated.

All requests for change will require a series of decisions to determine if a change is worth considering, if resources should be directed to assess the impact of a change, if the change should be implemented, and who needs to provide the appropriate resources and approvals.

Establishing a Change Management System

Establishing a change management system that harmonizes a process, actions, tools, owners, and their interactions in managing project change calls for a thoughtful system that will likely need to be tailored for each organization. The effectiveness of the Change Management System is dependent upon sufficient information about the project baselines, clear authority levels, and a good change management process.

Within the change management process there are different levels of authority. For example, a small change that will not impact the other aspects of the project can be approved by the project manager. However, for changes that may impact a baseline, resources, or more than one deliverable should go through the Change Control Board (CCB). Large changes that would entail re-baselining or would impact a project objective would need to be approved by the Sponsor and the Customer. The decision authority levels—project level, sponsor level, and customer level—should be documented to ensure a transparency and to avoid confusion.

Change Management Process

The change management process begins with a change request. The CCB is responsible for receiving and evaluating change requests and determining whether to accept, defer, or reject the proposed change. The CCB typically includes key stakeholders such as the project manager, customer, project sponsor, and technical leads. Of course, the needs of the project determine who should be on the board. A typical change management process is shown in Figure 12.1.

The process comprises the following steps:

1. *Change request.* A change request can be submitted by any valid stakeholder. A change request may be submitted with a paper form or, more often, in an electronic format. However, the request must be in writing.
2. *Change Log.* All changes must be documented in a Change Log or Change Register. The change log is used to record the ID, category, description of the change, requestor, date, status, and disposition.
3. *Analysis.* The CCB members, either individually or as a group, analyze the requested change. This includes evaluating the necessity, the cost, benefit, risk, and impact on the project, operations, and other projects.

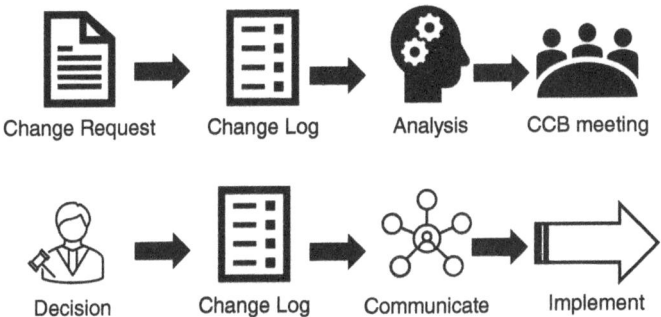

Figure 12.1: Example Change Management Process

4. *CCB Meeting.* At a Change Control Board Meeting all new and open change requests are discussed. The request is either accepted, deferred, or rejected. A deferral is used if more information is needed before they can make a decision.
5. *Decision.* Ideally the Board comes to consensus about whether to accept, defer, or reject the change. If consensus cannot be reached, then the board should follow its decision-making process.
6. *Change Log.* The Change Log is updated with the status of the Change Request.
7. *Communicate.* The decisions from the CCB are communicated to impacted and interested parties.
8. *Implement.* If the change was approved, there may be several documents and plans that need to be updated.

The change management process described can either be made more robust, or leaned out, to fit a particular organizational or project need.

Using the Change Management System

A Change Management System is especially vital in large or complex projects where there is a stronger need for change control. Because of their lesser scope and complexity, smaller projects can utilize a simplified change management process—the project manager often may be the only person available to handle change requests. In circumstances like this, an informal, but well-understood Change Management System would be more logical than having a formal system.

Senior management has a significant impact on the success of the Change Management System. They must set expectations and ensure stakeholders don't try to circumvent the system. For instance, it is not uncommon for a disgruntled change requestor whose change request has been denied to escalate to a senior manager and get an override to the decision. It is also not uncommon for a senior manager to attempt to mandate a scope change. Both examples demonstrate a lack of commitment on the part of senior management, and weaken the empowerment of the change management system, change control board, and the project manager.

The Change Management System's value is in bringing order to the project change process. Through a methodical prescription of the sequence and arrangement of roles, responsibilities, tasks, and tools involved in the system, this order significantly diminishes the possibility of problems, including scope creep, budget overruns, and schedule slippage.

CHANGE REQUEST

A Change Request enables the team to make conscious decisions instead of letting change happen in an ad hoc manner. Naturally, the change decisions tend to be of a higher quality. When the decisions are based on or coordinated with other tools necessary to assess the change impact, project scope, cost, and schedule are kept in check and in alignment with one another.

Most change requests are a result of several factors:

- *Value enhancements.* Changes that will increase the value proposition of the project
- *External events.* Changes caused by the environment in which a project operates such as new laws and regulations put into place.
- *Errors or omissions.* Changes that result due to forgotten or unrealized features or work during project planning.
- *Evolving scope.* Changes inherent in projects that don't start with well-defined scope (see "Scope Creep by Design")
- *Risk response.* Changes identified as a result of necessary risk mitigation or elimination responses.

However, the impact of changes on a project's schedule, cost, quality, and other matters may not be readily apparent to the person requesting the change.[2] As a consequence, the project may suffer. For this reason, it is very important to ensure that the benefit and value of each change is understood, and the impact then evaluated in a disciplined and professional manner before performing a change.

The Change Request is the first tool in implementing the Change Management Process. The content of the Change Request helps the Change Control Board (CCB) make decisions of a higher quality, keeping project scope, cost, and schedule in check and in alignment with one another.

Scope Creep by Design

Scope creep, or uncontrolled change of scope, is often perceived as a major threat to projects[3]. But one company faced with highly uncertain semiconductor fab projects controls the creep one change at a time. At the time of defining the scope historically prone to many changes, the company identifies a bucket of money equal to 10% of the project budget that is called AFC (allowance for change). Its purpose is to pay for the scope items that can't be predicted. To bring control to the process, every time such an item emerges, it is treated as a scope change and the project manager has to formally approve it. This very successful practice has helped the company to proactively manage scope creep from a budget perspective.

Developing a Change Request

Project Change Request forms come in many different styles and formats. When developing a Change Request for your project or organization, use the change management process contained in the Change Management System to guide the design and content to be included in the request form. Generally, there are five content sections that we recommend including in the Change Request, as shown in Figure 12.2:

1. Change request identification
2. Change request detail
3. Impact assessment
4. Alternatives
5. Recommendations and approvals.

Change Request Identification

This first section of the Change Request focuses on the quick identifiers for each requested scope change. Included in this section should be a concise change request title, a unique identification number, the name of the person originating the change request, and the date on which the change request is first submitted to the project change authority for consideration.

Change Request Detail

This section is intended to provide sufficient detail about a requested change so the project change authority can decide if there is sufficient cause for an impact assessment to be performed. The description of the change request must be sufficiently precise to provide clear understanding of which feature, deliverable, or work package will be changed and in what manner.

Additionally, this section must contain the justification for expending resource time and potentially changing the project baseline in order to incorporate the change. Justification should be articulated by describing the business benefits that will result from the change, if it is implemented. From the opposite perspective, we recommend that a well-formed justification also include a statement of the impact if the change is *not approved* and implemented.

Finally, the change requester should be required to make a determination of the criticality of a requested change. As shown in Figure 12.2, we typically use a 4-point

Change Request Identification	
Change Title:	
Change Number:	
Originator:	
Submittal Date:	

Change Request Detail	
Description:	
Business Benefit:	
Implications of No Change:	

Priority:
- [ ] 1 – critical: "project cannot move forward unless this change is made"
- [ ] 2 – high: "project success criteria impacted without this change"
- [ ] 3 – normal: "project value proposition enhanced with this change"
- [ ] 4 – low: "improvement in ease of use or performance gained by this change"

Impact Assessment	
Assessment Owner:	
Cost:	
Schedule:	
Resources:	
Deliverables:	
Impact Summary:	

Alternatives	
Alternative Approach:	

Recommendation and Approvals	
Recommendation:	

Decision:
- [ ] Approved
- [ ] Rejected
- [ ] Deferred

Figure 12.2: Project Change Request Template

scale of criticality that ranges from low priority to critical priority. A "Critical" change informs the CCB that the requested change requires urgent consideration and approval (see "Fast-tracking the Change Process"). It is of utmost importance that critical changes should not be an excuse to circumvent evaluation of matters related to quality, performance, reliability, safety, or any other aspect. Building safeguards that enforce appropriate consideration of the change may be a useful aid.

Fast-tracking the Change Process

ODI Incorporated's primary customer made it very clear that they had become frustrated with slow change request responses when their design chief made the following comment: "We are not willing to put up with your long turnaround time of our major change requests!"

Keen to retain customers, ODI redesigned its change management procedure, adding three vital changes. First, a rule was made that turnaround time for major customer requests will be 48 hours. These changes required a significant evaluation effort, including involvement of design, tooling, and manufacturing engineers, as well as marketing and purchasing experts. In addition, these people and their representatives were not collocated, so communication among them was time consuming and slow. Second, in response, ODI built an intranet site that significantly sped up the communication. Third, instead of using consensus decision making on the change board, typically a slow method, one of the board members was nominated the approver while others were considered reviewers and inputters only. The system redesign led to a drastic improvement, helping ODI to reestablish good relationship with this very important customer.

With the appropriate justification information in hand, and a sense of change request urgency, the project change authority now has what it needs to decide if an impact assessment is warranted, and if a change request needs to be fast-tracked. If so, an assessment owner is assigned, and he or she is responsible for filling out the next section.

Impact Assessment

Changes come in all sizes. For changes with minor, if any, impact, you can treat it as a minor corrective action, not impacting baseline scope, cost, schedule, and resources. In contrast, a major change can impact the scope of work, funding, or schedule requirements. These changes may warrant a replanning (or re-baselining) effort, including changes to the WBS, schedules, budget, and resource allocation. To make all involved fully aware of such consequences, spend adequate effort to assess the impact of the change on scope, quality, cost, and schedule and document it in the Impact Assessment section of the Change Request.

It is difficult to evaluate the impact of changes on schedules without a good Gantt chart or network diagram, as this is where dependencies between activities are shown. The dependencies help analyze how a change to one deliverable and its activities will affect dependent activities down the road. Still, more often than not, that is what occurs—a schedule impact assessment is made on the basis of a gut feeling.

To safeguard against risks related to such assessments, rely on network schedules to produce reliable estimates, even if you are dealing with a small project.

Requiring a cost estimate for the proposed change is a well-meaning strategy to prevent cost surprises. That humans have a tendency to underestimate the cost is well documented in many books and papers. Several decades ago as well as today, missing an estimate by 20+ percent is not unusual. What is unusual is the failure on management's part to take this tendency into account when evaluating a change request. Asking for a detailed estimate when the change request is proposed is a sound management safeguard against this tendency. If the change is major, it is possible to go further and request an estimate from an independent source to compare with the estimate of the change originator.

Even though a project manager may have budget and schedule contingency to utilize to implement a scope change, availability of resources to do the work may be constrained. For full evaluation of impact, a discussion concerning which resources are needed and for how long is necessary.

Lastly, a thorough impact analysis should also include information about changes to project deliverables and the final project outcome. Will any project deliverable be impacted by the change? If so, how? Will the quality of the project outcome be affected and how? These are just a few of the critical questions that need to be asked and answered to complete the analysis.

Alternatives

Some Change Request forms include a section requiring a statement on alternative solutions to the stated need. We feel this is a good practice, and as such, we have included this field in our example Change Request.

This requires the team to spend time in the *possibility space,* thinking about alternative ways to implement the change. In practice, it is sometimes the case where a simpler and lower impact solution is found than the original path that was being pursued. This is where engaging the project specialists in the evaluation process pays off.

Recommendation and Approvals

Armed with an impact analysis and alternative solutions, the final pieces of information needed in a well-designed Change Request are the recommended direction and documentation of the final decision. We include three decision alternatives to consider: (1) Reject the change; (2) Approve the change; and (3) Defer the change to a later date.

Using the Change Request

As conventional wisdom goes, a change should be proposed as soon as it is needed. We would like to add a little bit more precision here. The early conceptual stage of scoping a project is the time when there is limited if any use a Project Change Request as a change management tool. With the scope still conceptual in nature, it is impractical to attempt to control changes. In the later stages of scope definition, however, it is practical to start using the Change Request. For example, for a new product development project that has not experienced the beginning of design work or has not completed any product specifications would not need to apply formal management of change requests. Rather, it would start using a Change Request for changes when they either constitute a departure from the agreed design specification or affect the work to be performed for product design.4 The Change Request would continue to be used from the point forward.

When do you stop applying the Change Request? Although the question may seem less than meaningful, there will come a time when any change may impede the project, potentially leading to costly waste and rework. An effective action in such situations would be to impose a scope freeze, a mandate that no changes will be considered unless an overriding reason exists. An example may be a customer-funded requirement to add a new safety feature to a product once it is in final testing and validation. The scope should be frozen before a project enters final test and verification activities.

The Change Request is quite efficient to use, only a few minutes is needed to enter information into the form. But that is only the mechanics, which will likely be preceded by a substantial analysis that is a function of the size and complexity of the proposed change. In a small-scale change, determining its scope, cost, and schedule consequences may take 15–30 minutes. At the other end of the spectrum are major changes to major projects, where a group of experts may spend a week or two to fully assess the requested change's impact on the project's business purpose and goals. Documenting changes reduces project participants' confusion, and results in better-controlled scope change, lower total cost, and fewer delays.

CHANGE LOG

A Change Log is used to record and coordinate the flow of project changes (see example Change Log in Figure 12.3). A Change Log records each change request and assigns it a unique identifier, making sure the decision about it—whether it has been approved or rejected by the change authority—is recorded as well. When a request is approved and the change implemented, that information also becomes part of a Change Log.

The Change Log provides good oversight of all requested, rejected, approved, deferred, and completed changes. Such clarity is bound to decrease the confusion about what changes have or have not occurred that is often present when a Change Log does not exist.

Developing a Change Log

For projects that are fraught with changes, the Change Log serves as a repository of change which provides good oversight of all requested, rejected, approved, deferred, and complete changes. The Change Log fosters an environment of transparency, offering fundamental but brief information on all changes associated with a project. If a simple design is used, all project stakeholders will be able to gain insight on any and all scope changes in a short amount of time.

CHANGE LOG						
Project Title: _____ Date Prepared: _____						
ID	Category	Description of Change	Requestor	Date	Status	Disposition
Page 1 of 1						

Figure 12.3: Example Change Log

Development of a Change Log begins with gathering information previously developed. The Change Management System provides full understanding of the project change rules necessary for ensuring the right information is contained in the log. The Change Request form provides the necessary information for the summary section of the Change Log, as well as additional information about the impact assessment.

If one were to perform a web search for sample Project Change Logs, many different formats would emerge—some complicated and over designed and some too simplified to provide meaningful information. We believe that a well-designed Change Log should provide information in four important areas:

1. A high-level summary of each change requested
2. The category of change
3. Current status of the requested change
4. The disposition or decision.

Developing a Change Log centers on creating a format which, when filled out, provides the necessary information in at least these four areas.

Change Summary

The first section of the Change Log should give the reader a quick overview of each change that has been requested. Summary information begins with a unique identifier for each requested change (usually as simple as a chronological numbering system) and a brief description of the change.

The summary information should be documented on the Change Request form and can be directly transferred to the change log. The Change Request form also includes the person who originated the request and the date that it was submitted.

Category

The Change Log should provide information on the category of change. Categories can be scope, schedule, cost, quality, or other categories as appropriate for the project. Categorizing changes can help with evaluating and implementing changes. If there are a lot of scope changes, it could indicate that the requirements or scope were not well defined at the start of the project. If there are too many, it might be useful to re-evaluate the scope and create a new scope baseline that is more realistic. The same can be said of the schedule and cost baselines as well.

It is a common practice to integrate change requests in batches rather than one at a time. For example, at the end of a phase or on a quarterly basis. Categorizing changes allows makes it easier to identify all the resource changes, cost changes, etc.

Status

It is good practice to provide a field in the Change Log that communicates the current status of each requested change. Current status may include a number of states that a change request may be in. Some example status states include "Received," "In Assessment," "Pending," or "Completed."

Disposition

Disposition refers to the decision about the requested change. A change may be accepted, in which case actions would be taken to implement the change and update relevant plans and documents. It may be deferred, indicating that more information or time is needed before making a decision, or it may be rejected. Change logs serve as the documented reference for all decisions associated with changing the project. They provide a historical record of approved project changes.

Attempt to keep the Change Log focused on the four essential pieces of information described above and avoid over designing the log format. As described in "It only Takes a Spreadsheet," exotic or enterprise-level tools are not required to develop a highly effective Change Log.

It Only Takes a Spreadsheet

Final accounting is a painful part of any contractual project, especially if there were a lot of approved project changes. It is even more painful when there is no project change log, as was the case in a project that approved several hundred changes over the two years the project was active. Because of a lack of change policy and several changes in the project manager position, the log was never established. Then, at the end of the project, it took the project team several months and thousands of dollars of their time to track down all requested, approved, and rejected changes to include in a final change control log requested by the client. Eventually, the team learned a very simple lesson the hard way: it only takes a spreadsheet to create a useful change log.

Using the Change Log

A Change Log is a dynamic document that is used as soon as the first Change Request is submitted and is updated with subsequent changes requests and as the status and disposition of changes occur. It should always be up-to-date with the latest project information. Fortunately, it only requires a short amount of time to enter information into the Change Log once it has been generated.

Use of the Change Log does not end with the logging of the change decision unless a change is rejected. For the changes that are approved, the log must remain open and in use until it can be verified that a change has been fully implemented. Although it sounds paradoxical, sloppy change coordination may fail to catch a change that was never requested, but implemented nonetheless (see "What Do You Do With an Unrequested Change?") Should this happen, the change is entered into the log and fully documented to provide a historical record of the change.

What Do You Do With an Unrequested Change?

A computer engineer with no prior experience in contract project work learned that the project's computer vendor went out of business. To address this he simply ordered better and more expensive equipment from another, more reputable vendor. After all, that's what he had done many times for his own company's internal needs.

Four months later when the equipment was delivered, the project manager was more than upset with the engineer. The project manager asked the very pointed question, "Why did you change computer specs without going through the change request procedure?" Even worse, the project manager refused to pay the price differential between the original and new equipment, saying "I have no money in the budget and to get it I have to go beg my chief financial officer."

The epilogue? After several months of frustration, the project client approved the change. The moral to the story: Having a change procedure in place means nothing unless you train people to use it.

The Change Log provides a complete historical record of all changes to a project, therefore providing documented evidence on why changes were required. Additionally, it provides a record of who approved each change and when. This information is many times a required deliverable in contractual-based projects.

SCOPE CONTROL DECISION CHECKLIST

Changes in scope will affect the tasks that are performed on a project, which has a direct effect on project resources, which in turn directly affect the project cost and schedule. Changes in project scope therefore create a cascading effect as depicted in Figure 12.4.

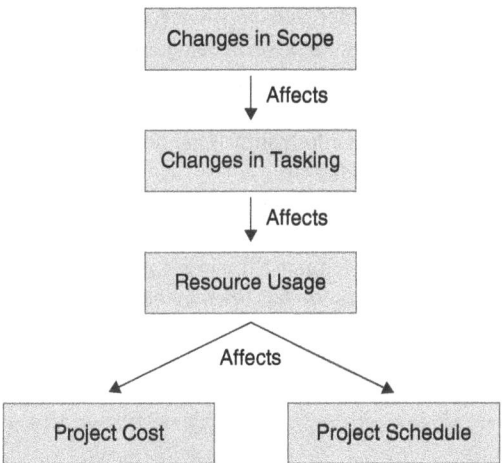

Figure 12.4: The Cascading Effect of Project Scope Change

The Scope Change Decision Checklist helps the project manager with decision quality and decision consistency.[4] High-quality decisions are those that are informed, reasoned, and thorough. Improving the quality of a decision begins with understanding the four primary steps in making a project scope change decision:

1. Description of the proposed change and benefits
2. Identification of the cost of change
3. Identification of solution options
4. Understanding the risks involved with the decision.

The questions contained in the Scope Change Decision Checklist are designed to improve the information used for decisions in each of the steps listed above.

Equally important to decision quality is decision consistency which involves making high-quality decision on a regular basis. Consistency in decision making comes with approaching decisions in the same manner over time. The Scope Change Decision Checklist provides consistency by documenting a standard set of questions to be asked and explored for each project scope change decision facing the project manager and his or her change authority.

Because of the cascading effects of scope changes, it is necessary to ensure that project scope management practices are effective.[5] Decision consistency is crucial for establishing and maintaining buy-in for the scope management process, and to prevent various project players from attempting to circumvent the decision process. A standard set of decision questions, formulated into a Scope Control Decision Checklist, is a valuable tool for a project manager to establish the level of decision consistency required on a project.

Developing a Scope Control Decision Checklist

A Scope Control Decision Checklist will be different for every organization because every organization will likely have a slightly unique change management system and process, complete with decision-making nuances.

Developing a standard set of decision questions for an organization is a good practice as it drives consistency across the projects being executed within the organization. The questions contained within the checklist can be developed by first understanding the change management process. Questions will center on what decisions have to be made as part of the process, what information is needed to make the various decisions, and what decision outcomes are desired.

Table 12.1 illustrates a sample set of questions that can be used as a reference for developing your own decision checklist, which should then become part of your PM Toolbox.

We have included a status column in the sample checklist that, if adopted, can be used to indicate whether the question has been asked and, if so, if appropriate action has been taken. Some project managers add a third column to the checklist labeled "documented" to indicate if the answer to the question has been documented.

Table 12.1: Sample Scope Control Decision Checklist	
Status	**Checklist Questions**
☑	Does the change strengthen the project objectives?
☑	Does the change negatively affect the project objectives?
☑	Does the change modify any of the original project objectives?
☑	Can the change be implemented within the current budget?
☑	Can the change be implemented within the current schedule?
☑	Do we need additional resources to implement the change?
☑	Does the change affect the customer?
☑	Have we considered alternative solutions?
☑	Does the change require senior management approval?
☑	What is the impact of not implementing the change?
☑	What base assumptions have been made concerning the change?
☑	What risks does the change introduce?
☑	What risks does the change reduce or eliminate?
☑	Do we need a contingency approach?
☑	How do we verify correct and complete implementation?
☑	Who is the final decision maker?

Using the Scope Control Decision Checklist

Once the initial scope of a project has been established, change will certainly begin to occur. Unmanaged scope change is a leading cause of project failure. For a good explanation see "Scope Change: If You Have to Do It, Do It Early."

Scope Change: If You Have to Do It, Do It Early

Uncontrolled project scope changes are known as project killers because they:

Cause delay.
Increase the cost.
Damage morale and productivity.
Spoil relationships among project participants.

Why such a far-reaching impact? First, more often than not, changes cause work to have to be repeated in the impacted activities. Second, any activities related to those directly impacted by an uncontrolled change will likely need to change as well. This means that the earlier you make the change, the less rework and less damage you inflict on the project. Early in the project, very few activities have been worked on and completed, so therefore there is less rework that will have to take place. In contrast, if a scope change comes in late in the project life cycle, significant and costly change may need to occur, and any work that has to be redone becomes wasted effort and cost.

Consider for example if a change comes in late on a product development project. The change may cause redesign and associated re-development, the repurchasing of factory tooling, fixtures, and materials, the remaking of prototype systems, and so forth. All this work has significant impact on the project. Even a seemingly insignificant change of a team member during project implementation may set a team back by a number of weeks or months. The lesson to learn here is to think hard and make changes early in the project cycle and put a very critical assessment eye on changes that come late in the project.

The decision checklist can be used to eliminate decision process and decision outcome ambiguity from the very beginning of the change management process. Obviously, not all answers to the questions contained within the checklist will come quickly and easily. Depending on the complexity associated with a requested change, it may take a number of days, or in some cases, weeks to discover the right information needed to make a decision. Use the checklist throughout project planning and execution as a guide and focusing mechanism to ensure effective, consistent, and expedient decisions are being made.

References

1   Heagney, J. (2022). *Fundamentals of Project Management*. HarperCollins Publishing.
2   Verzuh, E. (2015). *The Fast Forward MBA in Project Management*. Hoboken, NJ: John Wiley & Sons.
3   Robertson, S. and James Robertson. Mastering the Requirements Process: Getting Requirements Right, 3rd Ed. (London, UK: Addison-Wesley Professional, 2012).
4.   Pinto, Jeffrey K. Project Management, 5th Ed. (Upper Saddle River, NJ: Prentice Hall, 2019).
5.   Andler, Nicolai. Tools for Project Management, Workshops and Consulting: A Must-have Compendium of Essential Tools and Techniques (Erlangen, Germany: Publicis Publishing, 2016).

13

AGILE PROJECT EXECUTION*

Agile is a form of adaptive project execution, heavily used in software development as an alternative to traditional approaches that emphasize the sequential or linear process starting from requirements gathering, planning, designing, code writing, testing, and implementation. Alternative to a sequential process, agile methodologies emphasize an iterative workflow and incremental delivery of project outcomes in short iterations.

Agile practitioners will tell you that Agile is a mindset, not a methodology. The mindset is anchored by four values (www.agilemanifesto.org):

Individuals and interactions over processes and tools

Working software over comprehensive documentation

Customer collaboration over contract negotiation

Responding to change over following a plan.

This is not to say that the items on the right have no value, they do. However, the items on the left are valued more than those on the right. These values guide all Agile frameworks, methodologies, and approaches. While Agile approaches were initially used primarily for software development, overtime many of the approaches have found their way into other types of projects. In fact, many project practitioners are using hybrid ways of working, where Agile and non-Agile (waterfall) tools and techniques are used to deliver value.

Key benefits of Agile include:

- *Being customer-centric*. In Agile projects the customer is part of the team. They articulate the needs, prioritize features, and observe demonstrations. By having ongoing customer input Agile projects deliver the features and functions that are most desired early in the project.
- *Flexibility and adaptability*. With Agile, customers have the opportunity to add work, reprioritize work, or eliminate work. This can occur due to changes in customer needs, market trends, and technology.

---

*Peerasit Patanakul, James Henry, Jeffrey A. Leach equally contributed.

Project Management Toolbox: Tools and Techniques for the Practicing Project Manager,
Third Edition. Cynthia Snyder Dionisio, and Russ J. Martinelli.
© 2025 John Wiley & Sons, Inc. Published 2025 by John Wiley & Sons, Inc.

- *Deliver value early.* A key focus of Agile is to deliver value quickly and ongoingly. Using shorter iterations of work to deliver business value earlier allows customers to achieve value incrementally throughout the project. This can give the customers the opportunity to monetize projects earlier.
- Reduce project risks by allowing the customers to see the tangible progress of the project. This enables the customers to fully understand what is being delivered and confirm that what is being built is the same as what is needed.

Common Agile Misconceptions

Many practitioners and customers have false expectations of agile practices that can lead to misunderstandings and tension between stakeholders. Below are some of the more common misconceptions.

Agile projects don't require planning. Agile projects plan throughout the project rather than just at the start. This allows greater responsiveness to changing needs and market adjustments.

Agile projects don't require documentation. Agile projects produce documentation that is necessary, but don't spend a lot of time and effort documenting detailed plans. Because Agile practices are responsive to change, creating documentation that may not be used is unproductive. Rather there is a "just enough and just in time" approach to documentation.

Agile projects are done faster. The focus of Agile is on shorter iterations of work effort to produce increments of value sooner. This does not necessarily equate to the project being completed faster.

Agile is only for software development. While Agile was created to address challenges with software development projects, it is now used in many different types of projects.

Scrum is by far the most widely used methodology, and therefore this content will align with Scrum language and approaches. Where applicable we will present generic terminology that can apply to hybrid projects or other Agile methodologies.

SCRUM BASICS

The Scrum framework provides a structure of roles, meetings, rules, and artifacts that the project teams utilize during software development. Within this framework, a project is broken down into self-organizing teams (scrums) of about 6–9 members. Each team focuses on a self-contained area of work. To develop software, a list of customer wants and needs is created, referred to as a product backlog (see Figure 13.1).

Scrum uses fixed-length iterations, called *Sprints*, which are typically 2–4 weeks long. While in a sprint, scrum teams are responsible for taking on a set of features from the backlog and developing a deployable product that is properly tested. The team is

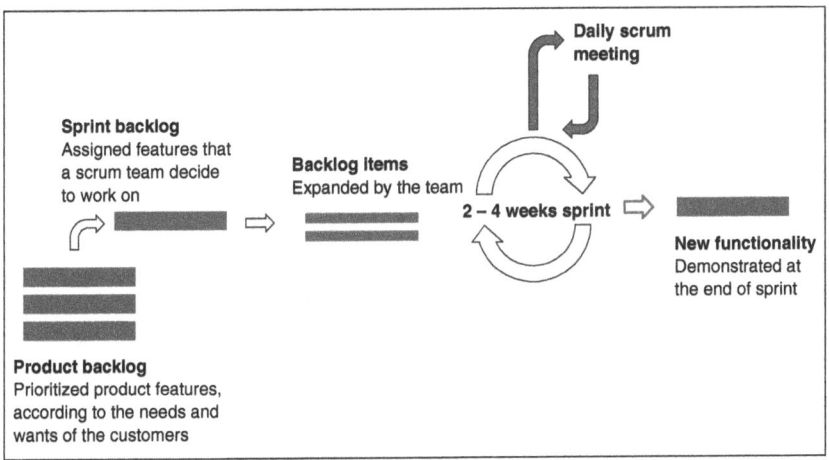

Figure 13.1: The Scrum Workflow

Source: Adapted from Eric J. Braude and Michael E. Bernstein. *Software Engineering: Modern Approaches*, 2nd Ed. (Wiley, 2016)

given full authority of how to successfully complete the sprint. Every day, an approximately 15-minute daily scrum meeting is organized to assess the status of the sprint, report on problems, and identify future tasks. At the end of a sprint, a customer demonstration is conducted. The leftover features and new tasks are gathered, the backlog is updated and reprioritized if necessary, and a new sprint begins.

This iterative approach allows the teams to develop a subset of high-value features as early as possible to incorporate early feedback from the customers. In essence, the Scrum framework emphasizes empirical feedback, self-organizing teams, and striving to build properly tested product increments within short iterations. Within the Scrum management framework, participants hold several key roles. They are Product Owner, Scrum Team, and Scrum Master.

The Product Owner represents the customer of the project. Based on the product requirements, the Product Owner provides customer-centric items (user stories) to the team, prioritizes the items, and adds, deletes, or reprioritizes items in the backlog. The Product Owner is accountable for ensuring the value of the project output to the business. New products will be promoted into production, only after the Product Owner accepts the output from the sprint demonstration.

The *Scrum Team* typically consists of 6–9 members across disciplines. These members are responsible for analyzing, designing, developing, testing, communicating, and documenting product increments at the end of each sprint. The team is generally self-organized but may have some interaction with the project management office depending on the organization's structure.

A *Scrum Master* is responsible for removing obstacles for the team to be able to deliver the project outcome. He or she is not the team manager. A Scrum Master helps ensure that the Scrum process is used as intended.

The following sections introduce six tools and techniques that are frequently utilized in a Scrum framework. They are the product backlog and sprint backlog, release planning, daily scrum meeting, sprint task board, sprint burn down chart, and sprint retrospective meeting.

PRODUCT BACKLOG AND SPRINT BACKLOG

There are two common types of backlogs that are used in the Scrum management framework. They are the Product Backlog and the Sprint Backlog. The Product Backlog is a prioritized listing of all items that may be developed at some point in the future. These items represent the customers' wants and needs. There is no commitment for when these items will be assigned to a team to be worked on.

A Sprint Backlog represents a list of items that a scrum team has taken from the Product Backlog and has committed to developing during the next sprint. A sprint may be known as an iteration, and the Sprint Backlog as an Iteration Backlog.

Information on the Backlogs

The Product Backlog is visible to any stakeholder and is composed of customer-centric features—descriptions of the desired product from the point of view of the customer or user. These could be in the form of user stories or use cases.[1] These features also have a level of priority assigned to them to create a forced-ranked list of desire functionalities. However, the priority of the items can be changed as needed. See Figure 13.2 for an example of a Product Backlog for a team that is developing a travel booking website.

Product Backlog	
Stories that may be worked	Priority
As a user, I'd like to sort flights by price so I can see the cheapest ones first.	1
As a user, I'd like to sort flights by number of stops & see non-stops first.	2
As a user, I'd like to sort flights by airline so I can see the carriers in alphabetical order.	3

Figure 13.2: Example Product Backlog

A Sprint Backlog contains elements from the backlog that will be worked on during the sprint along with the tasks needed to deliver the backlog items. Items on the sprint backlog do not need to be prioritized because they have been committed to for a particular sprint.

The Sprint Backlog is visible to the project team such that they can use it as a reference during daily scrum meetings. Figure 13.3 illustrates an example of a sprint backlog if the team decided to work on the first and third stories from the product backlog shown in Figure 13.2.

Populating Backlogs

The Product Backlog is owned by the Product Owner who prioritizes the items or adds and removes items from the backlog. Product Backlog Items (PBIs) are often written in a user story format with more emphasis on the "what" than on the "how" of a desired feature. For example, a story that may be written for a team that is developing a travel booking website, could be: "As a user, I'd like to sort flights by price so that I can see the cheapest ones first."

Well-written user stories will usually conform to the following common format:

As a <who>, I want to <what>, so that <why>

The Product Backlog goes through a refinement process to add or remove PBIs, reprioritize work, and further decompose PBIs that are too large and not clearly defined. This process can also include providing necessary technical information to help the team estimate the amount of effort to complete the PBIs.

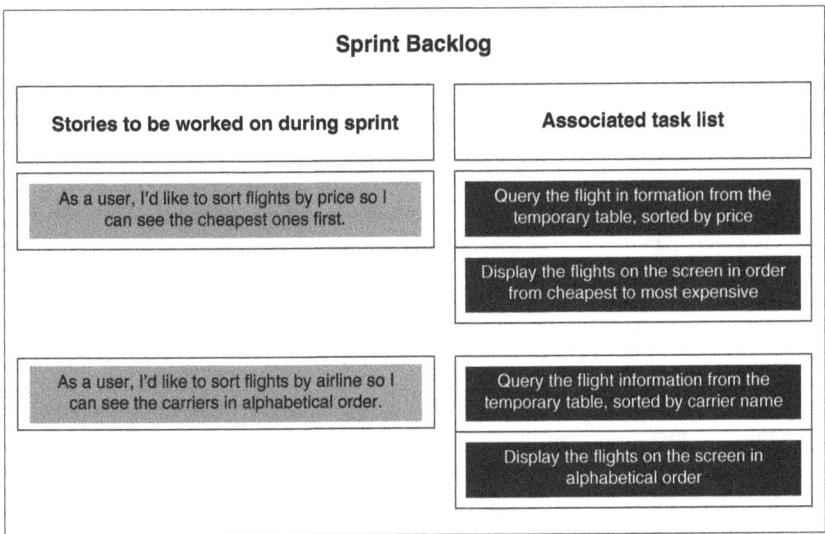

Figure 13.3: Example Sprint Backlog

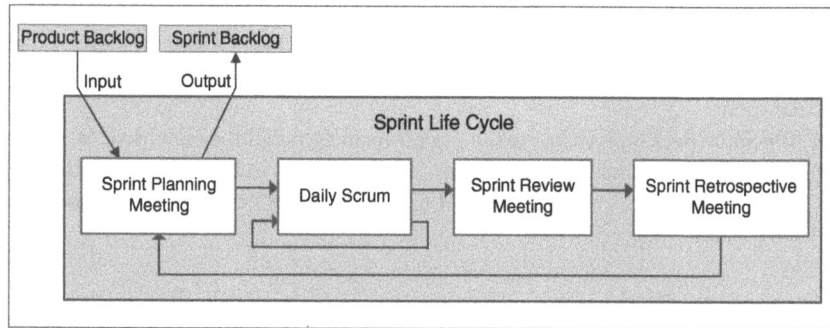

Figure 13.4: Product Backlog and Sprint Backlog in Sprint Life Cycle

Sprint Backlogs are populated by selecting the highest-priority items on the Product Backlog. There are instances where constraints prevent a high-priority Product Backlog item from being added to the Sprint Backlog. For example, some dependencies may need to be satisfied before the work can be started. It is for this reason that the highest-priority items are not always added to the current Sprint Backlog. However, it is important to note that the Sprint Backlog usually contains the highest-priority items from the Product Backlog.

The Sprint Backlog is created during sprint planning sessions. All team members, including the Product Owner and the Scrum Master, work together to agree to what will be worked on during the next sprint. The team considers their capacity, the constraints to development, and the priority of items when creating the backlog. Figure 13.4 shows the relationship between the Product Backlog, Sprint Backlog, and sprint planning meeting.

The Product Backlog allows for a close alignment between what the project teams are working on and what the business sees as the most important items for delivering business value. The ability to list everything that is desired along with the ability to prioritize, and reprioritize the work as necessary, enables the business to have more strategic agility to meet the challenges that they face.

The creation of the Sprint Backlog necessitates engagement and collaboration between the business and project teams. It results in a common understanding between all parties on what the team has committed to delivering.

RELEASE PLANNING

A release is a set of sprints. They are used to group sprints into a release. A release can be to the customer, consumer, or a system. Release planning gives the scrum teams the ability to plan some longer-range activities as the number of sprints in a release is typically 6–8 sprints.

Release planning becomes a valuable technique when large systems with interdependencies are being developed by multiple teams. In developing large systems,

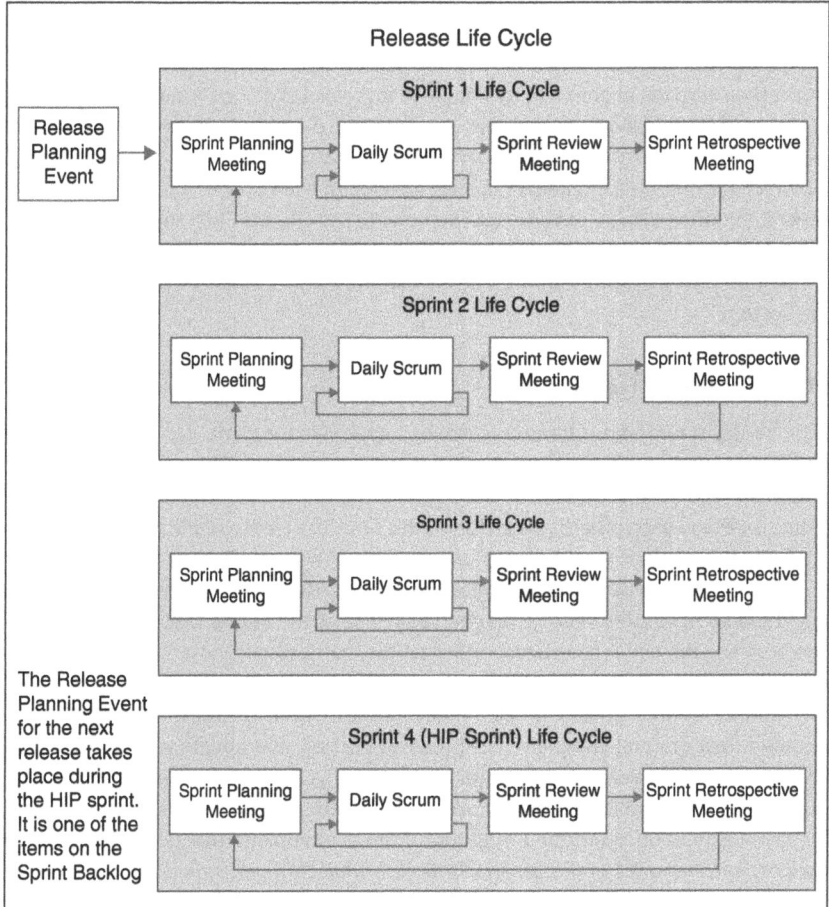

Figure 13.5: The Release Planning Event

release planning enables project managers and other stakeholders to have greater visibility into the expected completion of future milestones. In addition, the release planning exercise is an opportunity for collaboration between project teams, application teams, and other stakeholders within the organization to occur.

The Release Planning Event

The release planning event involves various stakeholders as it is typical for developing large systems. Often, project management and business leaders, as well as all the members of the impacted sprint teams are in attendance. The event is typically scheduled for two full days. For the first release, the release planning event takes place at the beginning of the release. For subsequent releases, the event takes place at the end of each release. Figure 13.5 shows the lifecycle of a 4-week release.

Initial Draft Release Plan

Before the inter-team collaboration can begin, each scrum team must work together to create their own initial draft release plan. This involves assigning Product Backlog items to the sprints included in the release. For example, if a team is involved in a 5-sprint release, they would consider dependencies and capacity when putting together a plan for which Product Backlog items they will develop in each of the next four sprints. The team would then display this plan for everyone to see (Figure 13.6). The last sprint is reserved as a HIP (Hardening, Innovation, and Planning) sprint. No Product Backlog items are planned for this sprint during release planning. HIP sprints are discussed later in this section.

Final Release Plans

After all the scrum teams have created their draft release plans, the plans are visibly displayed either physically or electronically in a way that allows for each scrum team the opportunity to review the release plans of all the other teams. Time is set aside for the teams to review each other's plans. During this time, the teams interact and are encouraged to ask questions to get a full understanding of what each team plans to work on during the next release. Through these discussions, each team member looks to find any dependencies between the planned activities of the teams. Once the dependencies between the teams have been identified, the teams discuss how to coordinate their activities. The draft release plans are then updated to account for any needed changes.

Time for another review of the release plans is given to the teams after they've updated their draft release plans. The process of review and update will continue until all the plans are believed to be coordinated properly. Before the release plans can be finalized, every member of every scrum team is asked to commit to the plan. If anyone, for any reason, is uncomfortable with any of the release plans, that person will explain their concerns and the entire group will discuss what changes should or should not be

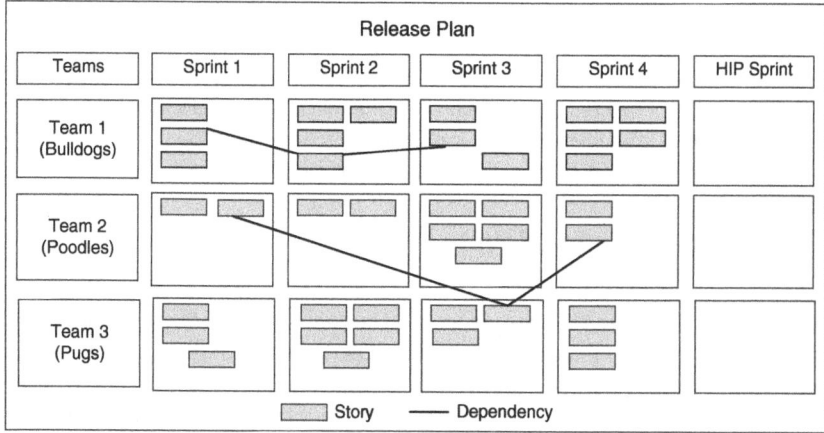

Figure 13.6: Release Plans Developed by Several Scrum Teams

made. The release planning event is not complete until everyone agrees to commit to the plan.

If your project does not entail multiple teams, the release planning can be less formal and won't take 2 full days.

Release Planning versus Sprint Planning

It is important to note that the release planning does not replace the sprint planning. At the beginning of each sprint, the Product Owner and team hold a sprint planning meeting to discuss and select the Product Backlog items the team will develop to create a working capability during the sprint. Traditionally there are two artifacts that result from a sprint planning meeting. The first artifacts are the sprint goals which are written by the team and describe what the team expects to get accomplished during the sprint. The second artifact is the Sprint Backlog which is the result of the sprint planning meeting.

As discussed earlier, while the Product Owner is responsible for indicating which items are the most important to the customers, the team is responsible for selecting the amount of work that they can implement. Toward the end of the sprint planning meeting, the team decomposes the selected items into an initial list of sprint tasks and makes the final commitment to perform the work.

While the focus of sprint planning is on each sprint, the focus of release planning is on the overall release consisting of multiple sprints. During some releases, the team could decide to move product backlog items into earlier releases because they have the capacity to do so. They may also decide to rearrange the order of items or even reduce their priority and drop them. Having identified the dependencies during release planning, the team can communicate these changes to the other teams as appropriate.

DAILY SCRUM MEETING

A Daily Scrum Meeting is a brief meeting that begins once the sprint planning meeting has been finalized and agreed upon by the entire team, see Figure 13.7. The primary structure for a Daily Scrum Meeting is to answer the following three questions.

1. What was completed yesterday?
2. What impediments were encountered that blocked me from being effective?
3. And finally what is planned for today or prior to the next scrum meeting?

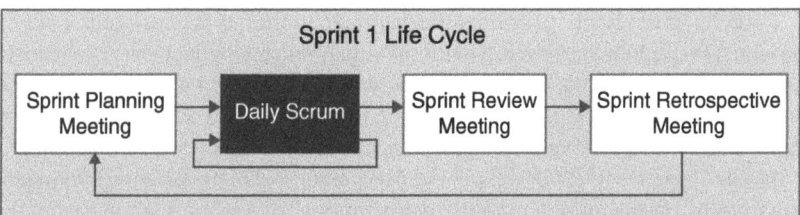

Figure 13.7: The Daily Scrum in the Sprint Lifecycle

By having each team member answer these questions, the team now has a complete picture of what's happening, the overall sprint progress toward its goals, and any modification that needs to be made to the upcoming day's work.

Within the daily scrum meeting the scrum team will leverage a task board as a way of communicating sprint progress. A task board, discussed later in this chapter, is very effective method of illustrating the current state of the Sprint Backlog over time. At a high level a task board contains columns representing user stories or Product Backlog items. A common set of columns is: user story, tasks to do, tasks in progress and task completed, see next section for more detail.

Organizing a Daily Scrum Meeting

A daily scrum meeting is typically held in the same location and time, each day. The meeting is most effective when conducted in the morning as it helps set the objectives for the upcoming day. The meeting usually lasts 15 minutes or less in which each team member will update the team on what was completed yesterday, impediments encountered, and finally what they plan to do for today.

Team members should focus on answering the three questions above with brevity in mind and avoid discussions not pertaining to the agenda. It is common that all team members stand up during the meeting, as sitting down has a tendency to extend the duration of the meeting; hence, the meeting is sometimes called a Daily Stand-Up. Any issues or concerns that require further discussion are noted and conducted after the Daily Scrum with the necessary stakeholders.

Participants

A Daily Scrum Meeting involves the entire scrum team as well as the Scrum Master. The Product Owner is considered optional and isn't required to attend every scrum meeting. In order to have an effective Daily Scrum Meeting it's important that all "participating" scrum team members come well prepared to quickly discuss the work they are responsible for completing.

It is important that the Scrum Master keeps the meeting on track and avoids any distractions from the meeting agenda. Meetings that stray from the agenda will more than likely run over the allotted time, begin to distract team members from answering their three questions, and even steer the meeting into sidebar discussions around fixing problems. When these types of distractions occur it's important that the Scrum Master brings the meeting back to focus as quickly as possible (see "Requirements for Effective Daily Scrum Meetings").

The Daily Scrum Meeting provides the team an opportunity for small course correction within each sprint as needed. It helps self-organizing teams get better over time by building trust amongst each team member. Similar to a sprint retrospective meeting (discussed later in the chapter), each meeting is based on the concepts of teamwork, empowerment, and collaboration through the use of open dialog and transparency.

A Daily Scrum Meeting is also a powerful technique for communicating information across a team. Having insight into each team member's area of focus allows each member to see the big picture while mitigating risk down the road.

Requirements for Effective Daily Scrum Meetings

#1-Resources: A daily scrum meeting requires that the scrum team be dedicated to attending the meeting. Dedicating so many resources to a daily meeting can negatively impact the overall velocity of a project if not managed properly.

#2-Preparation and Focus: A lack of preparation or attendance to the daily scrum meeting can definitely impact the overall value of the meeting. Team members should be prepared to answer the primary three questions below while dedicating the necessary time after each meeting to address any gaps or concerns

What work was completed the previous day?
What work will be proposed for today?
Did the team face any problems or impediments?

#3-Time: A Daily Scrum Meeting needs to adhere to the timeslot allotted (15 minutes or less). If meetings are running late on a regular basis, the Scrum Master should step in to ensure each team member is sticking to the defined agenda. This can be done by using side bar meetings after the scrum meeting to address any areas that may be pushing the meeting over the allotted time.

SPRINT TASK BOARD

For a scrum team, a Sprint Task Board is used to organize tasks into categories based upon their stage of completion. Sticky notes, with tasks written on them, are typically placed on the board to give an easy visual representation of the work being done and the progress made toward completing those tasks. The task board is a variation of a "Kanban Board."

A very simple example of a task board could have four columns labeled "User story," "Tasks To Do," "Tasks In Progress," and "Tasks Completed" (see Figure 13.8). At the beginning of a sprint, all of the tasks would be in the "To Do" column. As soon as a team member begins to work on a task, the sticky note with that task would be moved to the "Tasks in Progress" column. Upon completion of the task, the sticky note would be moved to the "Tasks Completed" column. At the end of a sprint, all the tasks should be in the "Tasks Completed" column.

Using the Sprint Task Board

Sprint Task Boards are not limited to only four columns as illustrated in Figure 13.8. Adding more meaningful columns could give sprint teams a more precise visual status. A task board may also contain different columns as befits the work. Figure 13.9 shows a

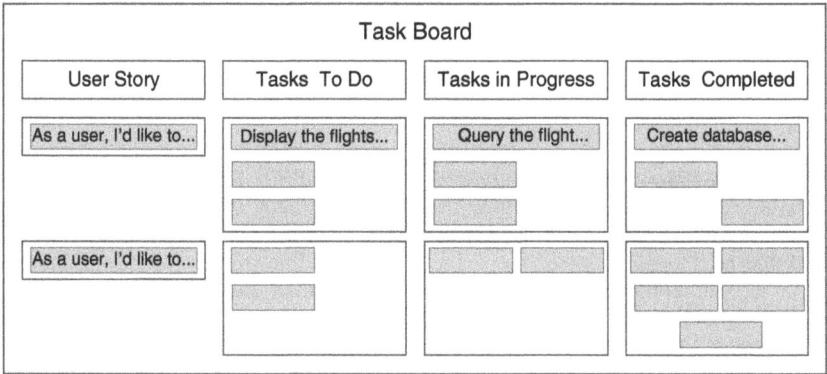

Figure 13.8: An example Sprint Task Board

Task Board					
User Story	Tasks To Do	In Design	In Coding	In Testing	Tasks Completed
As a user, I'd like to...	Display the flights...			Query the flight...	Create database...
As a user, I'd like to...					

Figure 13.9: A variation of Sprint Task Board

Sprint Task Board with columns for "Tasks to Do," "In Design," "In Coding," "In Testing," and "Tasks Completed."

It is also recommended that dependencies between tasks be noted on the sticky notes and that the team member working on the task adds his or her name as the task is moved to new columns. This will allow for easier communication and collaboration when the need arises. As illustrated in Figure 13.10, the task "Display the flights on the screen in order from cheapest to most expensive" is a successor of "Query the flight information from the temporary table, sorted by price." Once the task began, Jeffrey Leach who was responsible for design added his name to the task. When the design was completed, James Henry who is responsible for coding added his name to the task and crossed out the name of the designer.

Task boards make the progress of the team visible to everyone and they facilitate team coordination. If the Sprint Task Board is used properly, everyone should know who is assigned to each task, dependencies between tasks are properly accounted for, and there should be no duplication of effort.

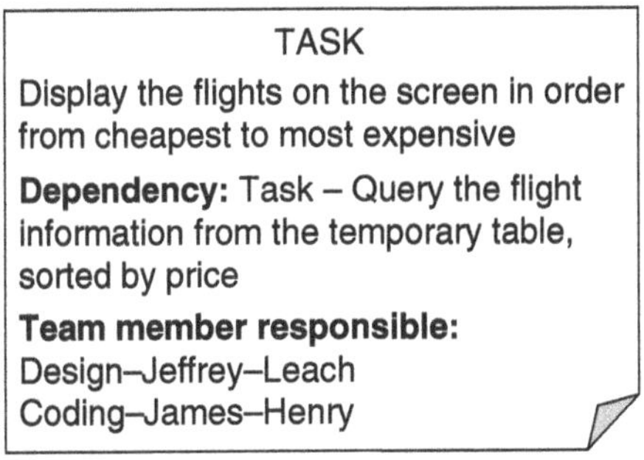

```
┌─────────────────────────────────────────┐
│                  TASK                     │
│ Display the flights on the screen in order│
│ from cheapest to most expensive           │
│ Dependency: Task – Query the flight       │
│ information from the temporary table,      │
│ sorted by price                           │
│ Team member responsible:                  │
│ Design–Jeffrey–Leach                      │
│ Coding–James–Henry                        │
└─────────────────────────────────────────┘
```

Figure 13.10: Sample Sticky Note for a Task

Additionally, a Sprint Task Board is a simple and easy way to manage the flow of work and information throughout a sprint. It visually represents each team member's work and progress over the duration of a sprint. By having the task board available for team members to consume it now allows team members to know what other team members are working on.

SPRINT BURN DOWN CHART

A Sprint Burn Down chart generally displays the total quantity of work or tasks needing to be completed versus the time allotted to complete a sprint (or an entire release consisting of several sprints). The chart is a simple and easy-to-use tool that provides relatively accurate estimates of the overall progress of a sprint or release progress.[2]

Having tasks broken down into sufficient detail is a necessary prerequisite for creating a Burn Down Chart. This is normally completed during the sprint planning phase where each task has been forecasted based on the hours it will take to complete the work. These estimates are normally determined by the entire team during the planning meeting.

Developing a Sprint Burn Down Chart

An example of the sprint burn down chart is shown in Figure 13.11. While the chart is rather simplistic, it is very beneficial to the team when utilizing it on a daily basis.

On the chart, the x-axis represents the amount of time that the team has defined for the entire sprint (20 days, in this example). The y-axis represents the total amount of time (in hours or days) the team will take to complete all of the given tasks for that sprint (120 hours, in this example). Since the amount of work to be complete will decrease over time, the general trend in the chart is to burn down to a point where zero work remains. A trend line can be calculated and drawn to illustrate when work might be completed.

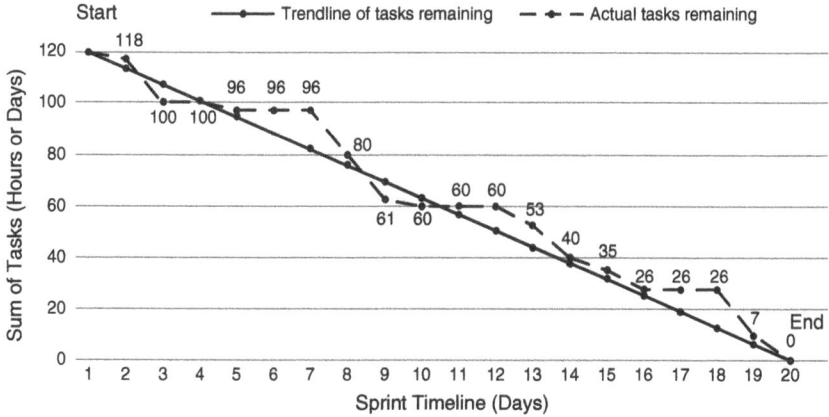

Figure 13.11: Example Sprint Burn Down Chart

Using a Sprint Burn Down Chart

A Sprint Burn Down chart is updated on a daily basis by each of the task owners. Collecting and communicating this information can be done by using many different techniques. Table 13.1 is a simple table that associates tasks to a user story along with a status, owner, projected hours, and remaining hours. In this scenario after spending an entire day working on Task B, the Business Analyst (BA) has determined that another 5 hours of work will be required to complete the tasking. The BA then updates this information using the "Remaining Hours" field located in the table.

After each task has been updated, the data is then aggregated and plotted for that day to determine whether or not the sprint was progressing as planned, see Figure 13.12. If the "Actual Task Remaining" line was above the "Trend Line of Task Remaining," this means the sprint is progressing at a slower pace than expected and may not complete the defined scope of work. On the other hand, if the "Actual Task Remaining" line was below the "Trend Line of Task Remaining," this means the sprint is moving at a faster rate than projected and may be able to finish ahead of schedule.

The Sprint Burn Down Chart provides each team member a visual representation of the total estimated time for all tasks to be completed (hours or days) versus the time

Table 13.1: Task Tracking Table						
User Story	**Task #**	**Task Description**	**Status**	**Owner**	**Projected Hrs.**	**Remaining Hrs.**
Story 1	A	Develop requirements	Closed	SME	8	0
	B	Coordinate with customers	In Work	BA	6	5
	C	Code module 1	Open	Developer 2	10	10

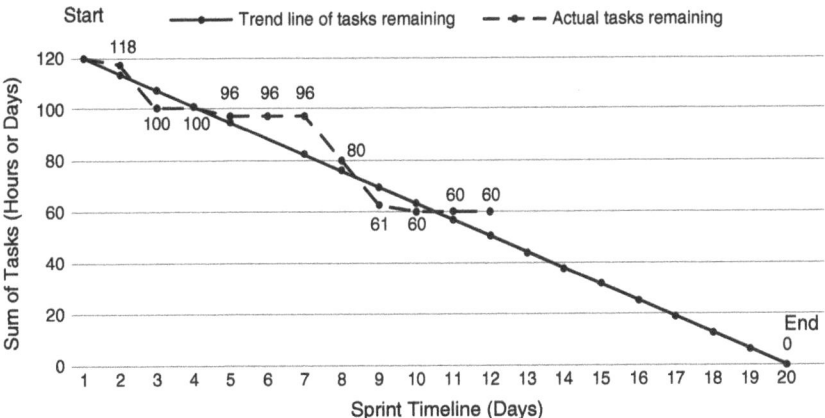

Figure 13.12: Plotting Work in Progress

permitted to complete a sprint or release. It allows team members to actively estimate the time it to complete each task. From there, data is aggregated so that team members can determine whether or not they are under or over executing across the sprint or release.

The chart communicates the overall status of the sprint or release to the entire team. If the project starts to deviate in a positive or negative fashion from the trend line the team will have this information to help mitigate potential risk down the road. Throughout this process, the chart facilitates effective communications across the team in a transparent fashion for all team members to consume.

SPRINT RETROSPECTIVE MEETING

A Sprint Retrospective Meeting follows a sprint review meeting. In the sprint review meeting stakeholders review the completed work. Thus, the Sprint Retrospective Meeting is the last step in the sprint lifecycle as shown in Figure 13.13.

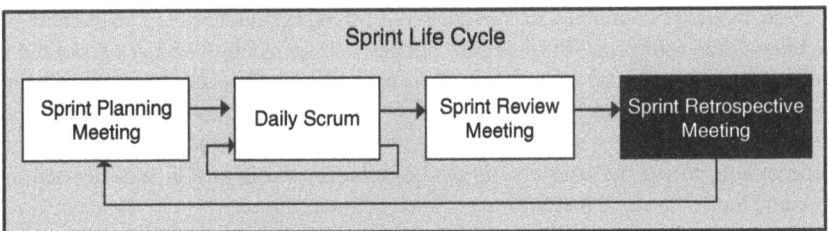

Figure 13.13: Sprint Retrospective Meeting in Sprint Lifecycle

The primary reason for a Sprint Retrospective Meeting is to provide the scrum team an opportunity to analyze and review the entire sprint with the intent of improving the overall process.[3] The meeting is based on the concepts of continuous improvement. At the end of each sprint a retrospective meeting is held to analyze and review all the things that went well, as well as the thing that didn't.

Within the meeting, the scrum team may choose to analyze existing processes, technologies, collaboration, communication techniques, and so forth. Once the Sprint Retrospective Meeting concludes, the scrum team will then carefully select the process improvement areas of interest and plan for them in the upcoming sprint.

Organizing a Sprint Retrospective Meeting

An effective Sprint Retrospective Meeting takes the necessary time to thoroughly discuss and document the sprint retrospective agenda. Typically, the duration of a Sprint Retrospective Meeting is about an hour and a half, but in many cases can require more time depending on the duration of the sprint, complexity of the project, the size of the team, or the team's overall experience with Scrum methodologies.

A Sprint Retrospective Meeting involves every scrum team member including the Scrum Master. In order to have an effective meeting, it is critical that all contributing team members feel comfortable and safe in their surroundings as they provide suggestions and recommendations on ways to improve the process.

In some cases, the Product Owner may not be involved in the Sprint Retrospective Meeting due to his or her potential decision-making authority (or power) over other team members. If the Product Owner does participate, they can comment on how the team is doing in maintaining alignment with the business goals and the customer perspective. However, their contributions need to be respectful and contribute to the feeling of trust on the team, not detract from it.

Using a Sprint Retrospective Meeting

The process of information gathering in the Sprint Retrospective can be done in many different ways.

Visual learning is a great technique that can help the Scrum Master facilitate communications across the team. As an example, a simple whiteboard can be divided into three areas: Good, Bad, and New (see Figure 13.14). Good items would identify areas that worked well over the sprint and didn't require any modifications. This would include things like continuing to limit daily sprint stand-up meetings to fifteen minutes or to continue using custom client side queries. Bad items on the other hand would include areas that did not work well resulting in process inefficiencies. Examples of this would include such items as defining the term "completed" or "done" more precisely for team members to better understand, to segment product backlog items into smaller components during the backlog refinement phase, or finally to shift the daily scrum meeting from 9:00 am to 8:30 am.

Finally, new items would focus on areas that hadn't been defined or were missing from the overall process. This would include such items as building a knowledge

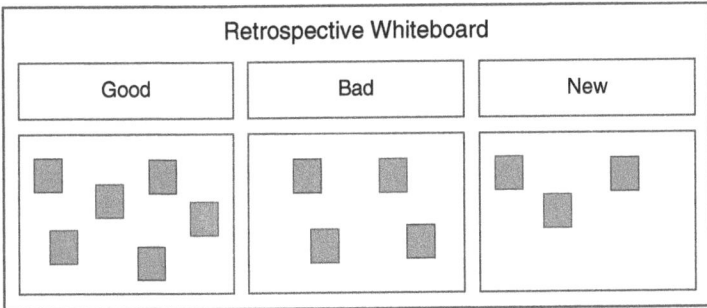

Figure 13.14: Retrospective Whiteboard

management system to inventory agenda items that fell outside the daily scrum meeting, or adding a video teleconferencing solution in the scrum room for external team members to enhance collaboration and communication.

There are many different techniques that a scrum team can utilize when gathering and categorizing information. As an example, sticky notes can be provided to each team member based on a color schema defined (Green = Good, Red = Bad, and Yellow = New). Each team member would then fill out as many sticky notes as needed in order to communicate their recommendations. Once completed each team member would place their sticky notes in a box for the Scrum Master to place on the whiteboard under the appropriate columns. Using a box ensures that all recommendations are anonymous and free from judgment, thus helping create a safe environment for team members to provide recommendations.

There are many other options for the Sprint Retrospectives, such as:

■ 4 Ls: Liked, Lacked, Learned, Longed For
■ Start doing, Stop doing, Keep doing
■ Anchors and Wind.

Many Scrum Masters switch retrospectives each time to keep the meeting productive and interesting.

Selecting Improvement Areas

Using a whiteboard such as the one illustrated in Figure 13.14, the team identifies and prioritizes which areas for improvement they'd like to focus on in the upcoming sprint. One popular technique is to provide each team member a few stars (3–5) which they would place on the items they feel would make the most significant impact to the project. If a team member felt very strongly about a particular area they could place all of their stars on it or they may choose to distribute their stars across several areas. Once all of the votes had been collected, the team would then tally the results to see which items received the most votes. From there, the output of the meeting would then be queued up for the Product Backlog meeting where the Scrum Master, scrum team, and Product

Owner would determine what items were most important and should be addressed in the next sprint.

Continuous improvement is critical to the overall integrity and foundation of a project as it moves throughout its project lifecycle. Overlooking continuous process improvement results in projects repeating past mistakes and impacting the overall efficiency of the project. Additionally, a Sprint Retrospective Meeting is based on the concepts of teamwork, empowerment, and collaboration through the use of open dialog and transparency. This is a powerful way to get each team member committed to the project as their feedback and recommendations are now being heard and applied across the project.

References

1.   Cohn, M. (2010). *Succeeding with Agile: Software Development Using Scrum*. Upper Saddle River, NJ: Pearson Education.

2.   Schwaber, K. (2004). *Agile Project Management with Scrum*. Redmond, WA: Microsoft Publishing.

3.   Schwaber, K. (2004). *Agile Project Management with Scrum*. Redmond, WA: Microsoft Publishing.

PART

V

Project Monitoring, Reporting and Closure Tools

14

SCHEDULE MANAGEMENT

S chedule management consists of assuring that project work is accomplished according to the planned timeline, assessing changes in the timeline based upon changes to the project, and establishing a new baseline schedule when necessary. Generally, there is minimal concern if work is completed earlier than planned, so primary attention is focused on preventing schedule slippage. The primary exception being if early work completion creates a cash flow problem later in the project or is due to taking quality shortcuts.

Schedule slips caused by scope changes or major resource adjustment are fairly easy to detect. However, humans are inherently poor at managing their time—both personal and professional (see "Tips for Better Time Management")—and as a result, most slippage occurs one day at a time and is more difficult to detect. Project managers need to be vigilant about preventing slippage from accumulating to a conspicuous and unacceptable level. To do so, they need to have tools that provide early warning of potential and realized schedule slips.

The tools presented in this chapter provide a variety of options for early warning detection of schedule slippage. Some tools are more suited for smaller projects, some for larger projects, others for complex projects, and still others for simple projects. Since the type of project a project manager will be called upon to manage will likely vary over the course of his or her career, it is recommended that each tool become a part of all project managers' PM Toolbox.

Although schedule slips are common, they are not inevitable. Given early detection, there is much a project manager can do to eliminate or limit them. For example, they can sub-divide tasks into smaller chunks of work, they can reassign resource responsibilities, they can add additional resources, different skill sets, or different skill levels, they can take advantage of schedule slack by adjusting task start/stop dates, and so on. If schedule slippage is detected early enough to enact control actions, project managers stand an excellent chance of managing to their baseline schedule and meeting their timeline commitments. Beginning with the Slip JChart, the following tools are designed to assist project managers perform their schedule management duties effectively.

Project Management Toolbox: Tools and Techniques for the Practicing Project Manager, Third Edition. Cynthia Snyder Dionisio, and Russ J. Martinelli.

Tips for Better Time Management

Better management of our time can help to improve our efficiency in both our personal lives and in our professional lives. The following ten tips for better time management are shared by Shirly McDowell, a seasoned project manager for a large financial institution.

1. *Prioritize.* Since it is impossible to do everything, learn to prioritize the important tasks and let go of the rest. This is done to avoid unwanted delay in the important tasks and at the same time avoid any chaos resulting from the delays.

2. *Know your deadlines.* When do you need to finish your tasks? Mark the deadlines clearly so you know when the work needs to be completed.

3. *Target to complete early.* With the deadlines identified for your tasks in the previous steps, plan your activities in a manner that will result in early completion if all goes as intended. This will leave room for the unexpected things that will inevitably occur.

4. *Know the intended results.* Make sure you understand exactly what it is you are trying to accomplish at the end of each task. This will help you know what success looks like before you start. It will also assist in knowing when to *stop* working a task.

5. *Create a daily plan.* Plan your day before it unfolds. The plan gives you a good overview of how the day will play out and helps focus the mind.

6. *Focus.* We are all guilty of excessive multi-tasking. Focus on one task at a time until it is completed or at the intended state you planned for the day. Focus solely on what needs to be done according to your plan in order to increase your efficiency.

7. *Plan for interruptions.* Plan to be pulled away from what you are doing—it's inevitable. This means not creating a daily plan that won't accommodate unexpected interruptions in your day. Allow for the interruptions instead by making them part of your daily plan.

8. *Learn to say no.* Don't take on more than you can handle. For additional tasks and activities that come your way when you are focusing on your high-priority tasks, give a firm "no" or defer them to a later time.

9. *Block out distractions.* It is important that you avoid any forms of distractions while working on the tasks that you have identified as high priority. When focusing on high-priority tasks turn off the phone, close your browser, close instant messaging, close your email. Don't give people your attention unless it's absolutely crucial in your business to offer an immediate human response. This will improve your concentration.

10. *Don't procrastinate.* There is no benefit of putting off tasks that you find uninteresting, difficult, or undesirable if they are important to accomplishing the work you have been assigned. Instead, try taking on these tasks early in the day when you are fresh and have the energy to focus on completing the tasks as quickly as possible.

SLIP CHART

The Slip Chart tracks progress of the overall project schedule by showing an estimate of how much time the project is ahead of or behind the baseline schedule at the time of reporting (see Figure 14.1). When updated on a regular, periodic basis, the Slip Chart provides a project manager and project stakeholders a good, overall view of schedule performance in relation to the plan. When consecutive estimates are linked, a trend line is formed. As a result, the Slip Chart can help predict the project completion date and signal the need for corrective actions to respond to potential slips in the completion date.

Developing the Slip Chart

Developing a Slip Chart begins with gathering the baseline schedule, current schedule performance, and all approved scope change requests that haven't been incorporated into the current baseline schedule.

It is a best practice to include the project team in this activity, especially team members who own critical activities and near-critical activities. The activity owners deconstruct activity dependencies to assess their impact on the progress of subsequent

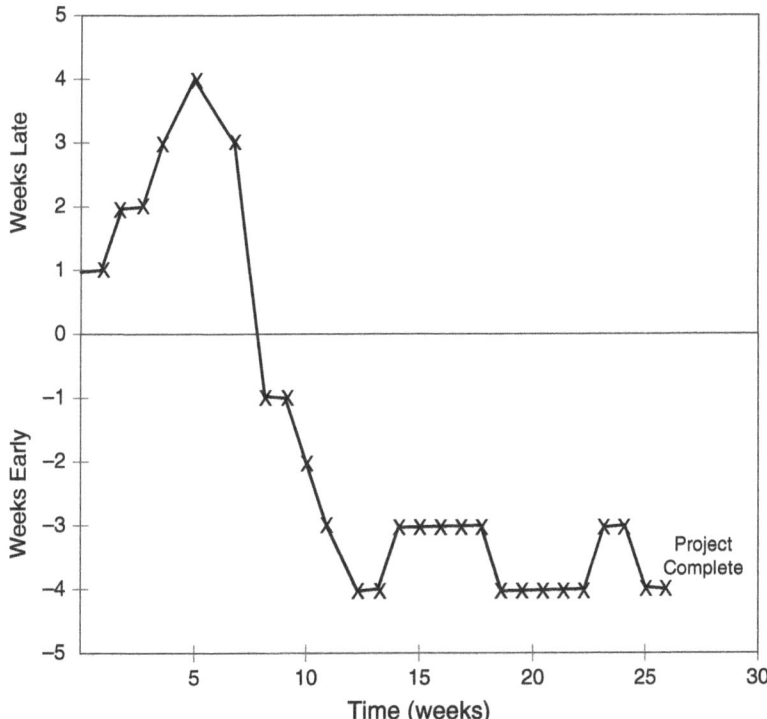

Figure 14.1: An Example Slip Chart

activities. Other owners join the impact analysis, receiving and giving more information to calculate how slips of individual activities on the critical path combine to establish the slip in the overall schedule.

Review Project Progress

For the first Slip Chart review meeting, create a template by laying the project timeline along the horizontal axis, "time zero" in the middle of the vertical axis, and early and late increments above and below the "time zero" point. Figure 14.1 illustrates the format of the horizontal and vertical axes, with the unit of time a variable that can be set to days, weeks, months, or quarters depending on the overall length of a particular project.

For subsequent Slip Chart review meetings, indicate the amount of slip associated with the critical path or critical chain, based upon current schedule performance. If it is determined that the schedule end date will be later than planned, go *up* the vertical axis by the appropriate time and mark an "*x*" along the current date. If the schedule end date is pulled in and will be earlier than planned, go *down* the vertical axis by the amount the project will complete early. Repeat on a regular basis and connect the periodic updates via a line that reveals an overall schedule performance trend.

Predicting Schedule Completion

As stated earlier, the primary utility of the Slip Chart is in the demonstration of historical and current schedule performance against the baseline plan. However, it can and is used by some project managers to predict the changes to the project end date. Since the Slip Chart focuses on the project critical path, positive or negative slips indicated on the Slip Chart can be carried forward to adjust the project completion date.

For example, in Figure 14.1 the Slip Chart indicates that the project is four weeks behind schedule when the project status is reviewed during week five. The project manager can then use this information to develop corrective action plans and, if necessary, to manage stakeholder expectations regarding late delivery. If the project manager uses the practice of updating the project end date each time a change in schedule slip is indicated on the Slip Chart, he or she risks causing frustration and losing credibility on the part of the project sponsor, stakeholders, and partners (see "The Window May Be Closed").

The Window May Be Closed

Soon after the start of a hardware development project, the project Slip Chart showed a three-week slip. The team added the slip to the project completion date, predicting the project would be three weeks late. Here is an example of how this extrapolation may be a risky practice. One of the project's later critical activities, a one-week-long rapid prototyping, was to be subcontracted to a vendor, who accepted the activity's start date with the comment, "If you come to us a week later, that's fine. If you come later than that, our window will be closed. At that time, add seven extra weeks to the planned delivery date for the prototype. We have already committed to starting another project at that time."

Apparently, the extrapolation is misleading. The predicted completion date is not three but at least seven weeks late. The learning here is to be careful with extrapolations and ensure schedule risk is fully comprehended.

Using the Slip Chart

Small and simple projects can benefit from the Slip Chart, as can large and complex projects. The Slip Chart's value is primarily in its ability to record the history of project progress and thus reveal the historical trend.[1] The information gleamed from the Slip Chart can be used as input to other schedule management tools with superior predictive capability if a project manager needs to estimate future schedule performance based on historical and current information.

The visual and simple nature of the Slip Chart makes it easy for the project team to create and for executives to understand the information it conveys.

BUFFER CHART

The Buffer Chart plays a very similar role for project managers using the critical chain scheduling as the Slip Chart does for critical path scheduling (Chapter 7). It measures the status and consumption of buffers established by the critical chain schedule methodology to provide an early warning system to protect the project's end date. First, the Buffer Chart takes an instantaneous snapshot of buffers' percent consumed relative to the percentage of the work completed on the critical chain (see Figure 14.2). Consecutive snapshots taken at regular periodic intervals are then linked on the chart to obtain a line indicating the trend. For example, in Figure 14.2 the line suggests that the buffer is being consumed at a faster pace than the pace of progress in completing the critical chain activities. In other words, the line answers the question "How are we doing today?" providing information to make a proactive decision to impact the value of the schedule buffer. For the project shown in Figure 14.2, that decision would be to initiate actions to recover the project buffer.

Developing the Buffer Chart

Developing a Buffer Chart begins with gathering the baseline schedule, current schedule performance, and all approved scope change requests that haven't been incorporated into the current baseline schedule.

Review Critical Chain Schedule Progress

Developing a Buffer Chart is most effectively accomplished through involvement of the critical leaders on a project. This normally involves the people responsible for the

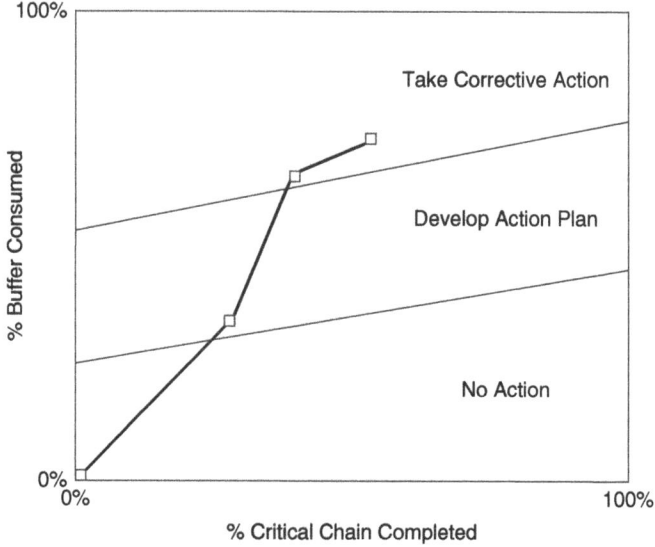

Figure 14.2: An Example Buffer Chart

primary tasks and associated deliverables. The critical chain schedule review meeting is focused on updates from those owning activities being executed on the critical chain, and secondarily on the subordinate merging paths.

Activity owners answer the crucial question of "How many days are remaining on this project activity?" While measuring the project health in this fashion is beneficial for overall good control of the project, it also helps with the next step—monitoring the buffers.

Knowing how many days remain on activities underway indicates the completion date for the activity. The date, then, provides background information to answer the next question, "What percentage of each buffer is consumed?" The emphasis is on the consumption of all buffers, including the project buffer and all critical chain feeding buffers on the merging paths. It is realistic to assume that the estimates may vary on a daily basis and may even go beyond their baseline duration estimates. As long as the activity owners continue sticking with critical chain scheduling principles of work and behavior, actual durations of their activities are of minimal concern.

Monitor the Completion of the Critical Chain

While the consumption of a buffer is of utmost significance, its consequences can only be understood in the context of the performance on the activity chain associated with the buffer. In the progress review meeting the team needs to estimate the percentage of work that is completed on critical chain and other activity chains. When this information is available, the team can compare the percentage of each buffer consumed with the percent complete of the activity chain associated with the buffer to establish the project's status or health at any given time. This comparison is conveniently illustrated on the Buffer Chart (Figure 14.2).

Create and Update the Buffer Chart

Prior to the first review meeting, the Buffer Chart template must be created. The horizontal axis represents the percent complete of the critical chain, from 0% to 100%. The vertical axis represents the amount of schedule buffer consumed, again from 0% to 100%.

The chart is then sectioned into three parts, representing the action needed relative to the combination of critical chain completed and buffer consumed. Following the guidelines established by Eliyahu Goldratt who pioneered the chart, we divided the Buffer Chart illustrated in Figure 14.2 into thirds.

Once the Buffer Chart is created, the progress review meetings focus on periodically updating the chart. To update, begin by marking the critical chain (or activity chain) percent completed on the horizontal axis at the time of the progress meeting. Go vertically upward until reaching a point equal to the percent of buffer consumed and connect this status point with point zero on the horizontal axis. Repeat drawing status points in each progress meeting, creating a line consisting of connected consecutive status points.

Using the Buffer Chart

Being an integral part of the critical chain schedule methodology, the Buffer Chart's use is closely linked to how the critical chain schedule is used. The purpose of the chart is to provide an anticipatory tool with clear decision criteria. Buffers, expressed in time units, are used to measure activity chain performance. The crucial point is to establish explicit action levels for decisions expressed in terms of the buffer size, measured in days. Goldratt, the developer of the Critical Chain scheduling method, proposes the following decision criteria.[2] If a buffer is negative—for example, the latest activity on the chain is late compared to its original completion date—and you penetrated the first third of the buffer, take "No action" (see the graph). Should you start consuming the middle third, it is time to assess the problem and develop a "Plan" of action. Once you are within the final third, you need to "Act." Note that these hold true for both the project buffer and critical chain feeding buffers.

For the Buffer Chart to be beneficial, it should be updated as frequently as one-third of the total buffer time. The reason is simple—the decision criteria are based on thirds of the buffer length. For example, whether a buffer is less or more than a third of the total buffer late (less or more than 5 days for a 15-day buffer) determines the type of action we take. In contrast, the chart in Figure 14.2 uses slightly different decision boundaries. See the example titled "You Need to Experiment" for an explanation.

You Need to Experiment

The Buffer Chart is based on the distinct philosophy of critical chain scheduling. To really comprehend its potential and put it to best possible use, you need to experiment with it and find the decision trigger comfort zone. For example, some companies modified the original criteria for using the buffers. Rather than relying on buffer consumption thirds as decision triggers given by the originator of the tool, they chose decision triggers that change as the consumption of the critical path changes. Decision triggers are borderlines between zones of "No action," "Plan," and "Act," mandating which action type to use. Look at the chart in Figure 14.2 and notice the subtlety of the slope of the decision trigger lines. The higher the percent of critical chain completed, the higher the decision trigger boundaries as a percentage of buffer consumed. This, of course, makes sense in general—the more work the project team completes, the more buffer consumption the project team can tolerate. But an exact amount of "More" should be picked by the company to fit its business purpose and nature of projects. Experiment to find the decision levels that best fit your company's projects.

JOGGING LINE

The Jogging Line is an excellent tool for showing the project schedule performance in aggregate form. It shows the amount of time each project task is ahead or behind the baseline schedule by drawing a line representing schedule performance in relation to the current date. In that manner, the line provides a snapshot in time indicating the fraction of work completed, and what remains to be completed (see Figure 14.3).

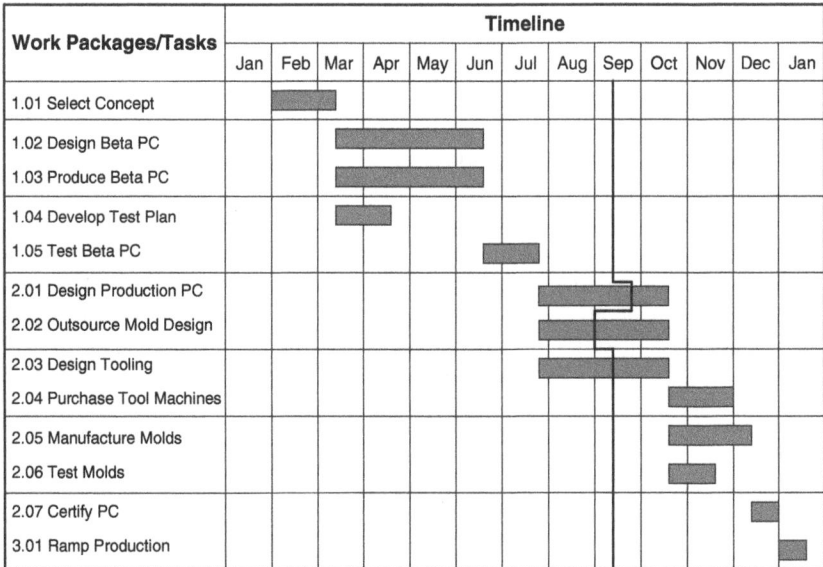

Work Packages/Tasks	Timeline												
	Jan	Feb	Mar	Apr	May	Jun	Jul	Aug	Sep	Oct	Nov	Dec	Jan
1.01 Select Concept		▓											
1.02 Design Beta PC			▓▓▓										
1.03 Produce Beta PC			▓▓▓										
1.04 Develop Test Plan			▓										
1.05 Test Beta PC						▓							
2.01 Design Production PC							▓▓						
2.02 Outsource Mold Design							▓▓▓						
2.03 Design Tooling							▓▓▓						
2.04 Purchase Tool Machines										▓			
2.05 Manufacture Molds										▓▓▓			
2.06 Test Molds										▓			
2.07 Certify PC												▓	
3.01 Ramp Production												▓	

Figure 14.3: An Example Jogging Line

The information gained by assessing the amount of time each task is ahead or behind the baseline schedule is used to predict the project completion date and create corrective actions necessary to eradicate any potential delay. Since the Jogging Line focuses on all project tasks and is created as a project team activity, it is best suited for use on smaller and less complex projects.

Constructing a Jogging Line

Constructing a Jogging Line begins with gathering the baseline schedule, current schedule performance, and all approved scope change requests that haven't been incorporated into the current baseline schedule.

Review Progress with Tasks Owners

As a first step, we suggest the project manager have a conversation with each activity (or work package) owner to gain a good understanding of the current status of the project work scheduled to be taking place at that point in time. As illustrated in Figure 14.3, not all project tasks will be of interest, as some will have already been completed and some are not scheduled to begin yet. For the tasks that are currently being worked on, the project manager should inquire about the following:

1. The progress of each task that the person is responsible for completing
2. Whether the current progress differs from the plan
3. If so, what issues led to it
4. When they anticipate finishing the tasks
5. If a variance to completion exists, what can be done to finish the tasks as originally planned.

Next ask how the answers to the task questions will translate into the activity progress—ahead or behind the plan, predicted completion date, and major corrective actions.

Essentially, this initial conversation is a rehearsal for the project progress meeting that will focus on constructing the Jogging Line for the project.

Review Project Progress

Assemble the task owners in a meeting along with the information gathered from the initial conversations with the task owners. Focus on asking the following five questions to gain a sense of overall project progress:

1. What is the variance between the baseline schedule and actual performance of each task?
2. What are the issues causing the variance?
3. What is the current trend—and current prediction of the completion date given the current performance?
4. What new risks have been discovered that may affect the predicted completion date?
5. What actions should be taken to bring performance back in line with the baseline (if a variance exists)?

For those familiar with the work of Deming, one will recognize that the five questions above are an integral part of the "Plan–Do–Study–Act" cycle (see "Project Assessment Questions and Deming's PDSA Cycle").

Where there are interfaces and dependencies between task owners, focusing on them is crucial since that is where the potential integration issues may appear on the project.

Project Assessment Questions and Deming's PDSA Cycle

The five questions of project assessment are in harmony with Deming's Plan–Do–Study–Act cycle, a circular approach to project performance improvement. Once the schedule baseline is established in the *Plan* step and project work is being carried out in the *Do* step, project assessment questions one, two, three, and four come into play in the *STUDY* step (see Figure 14.4). Here the schedule variance is established, its cause determined, and trend forecasted based on current issues and future risks. Then, in the *ACT* step, question five leads to the identification of connective actions, which will be planned for and implemented in the plan–do steps of the next project performance assessment cycle.

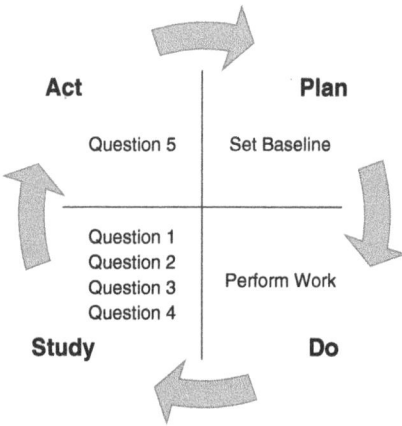

Figure 14.4: Project Assessment as Part of the Plan–Do–Study–Act Cycle

Draw the Jogging Line

Now is the time to assess the current progress of the project by drawing the Jogging Line based upon performance status information from each of the task owners. Begin the exercise by showing a plot of the baseline schedule that shows the critical path.

Next, mark the date of the progress meeting, whether it's called data date or the reporting date. From that date, draw a line vertically down until reaching the first task or work package that is currently being worked on. The task owner will tell how many days the work is ahead or behind the baseline. It is of vital importance that their information is reliable. If that's not the case, see what may happen in the example titled, "Are You Getting Accurate Information?"

Draw a horizontal line to the left of the data date for as many days the actual schedule is behind. Or, if you are ahead of the baseline schedule, draw the horizontal line to the right of the data date for as many days. At that point, draw a vertical line crossing the first activity. Repeat this exercise for all other tasks being worked on at that time. When vertically crossing the last activity, draw the line horizontally back to the data date and then turn vertically downward. In this way, the Jogging Line begins from and ends at the data date.

Are You Getting Accurate Information?

Pamela, a project manager for a major appliance manufacturer, is struggling with inconsistent progress reporting information from one of her team leaders. According to Pamela, "In a progress meeting, Jim told me his deliverable was three weeks behind schedule. A week later, he reported that he caught up with the deliverable and was right on schedule. Knowing that only one person was working on the activity, I polled several experts and asked how many hours it would take to catch up on the work. The answer was about 140 hours. That means his engineer would have to have worked about 180 hours in the past week."

Pamela is clearly receiving inaccurate status information, which is delivered verbally on a weekly basis. This is a scenario where another tool should be used to force more accurate status reporting from within the project team.

Predict Project Completion Date

As shown in Figure 14.3, work package 2.01 is ahead of schedule by approximately a week, while work package 2.02 is behind schedule by about the same amount. If either of these tasks are on the project critical path, an adjustment to the project completion date, as well as the timing of downstream dependent tasks, will have to be made.

Knowing how much each activity is ahead of or behind the schedule prepares you for making an educated forecast of the project completion date. If changes to the baseline schedule result, the next discussion in the progress meeting has to be about actions that can be put into effect to finish as originally planned. It is best to have the initial discussion about corrective actions with the task or work package owners (instead of doing it alone) in order to gain multiple perspectives of options and constraints. The project manager can then assess that information to make a decision on actions to put into play.

Using the Jogging Line

Small and less complex projects can use a manual version of the jogging line. For larger or more complex projects the % complete is entered into the scheduling software. Then select the option to show the jogging line.

A reasonably skilled and prepared project team can prepare a Jogging Line for a 25-task project in 15–30 minutes. As the number of tasks or work packages grows, so does the necessary time to construct and update the tool.

When constructing a Jogging Line, the following guidelines should be used:

- The line should be continuous
- It should start at the data date
- It should cross each task to indicate its time variance
- It should end at the data date.

Keep in mind that the Jogging Line only provides a snapshot in time of project performance and must be periodically updated.

MILESTONE PREDICTION CHART

The Milestone Prediction Chart anticipates the expected rate of future project progress by focusing on major project events—milestones, major deliverables, and project completion. Figure 14.5 illustrates an example of a Milestone Prediction Chart. Note that the vertical axis shows the team's predicted completion date for a specific milestone or deliverable, while the horizontal axis shows the date the prediction was made.

Obviously, the beginning point on the horizontal axis is the time when the schedule baseline is prepared, and its milestone dates are marked on the vertical axis. Once the project work is kicked off, the team reviews progress regularly and makes milestone predictions. By connecting all predictions for a particular milestone into a line, we can obtain the milestone trend line. If the line approaches the completion line moving upward, the trend would indicate a slip in the milestone or deliverable completion date. Delivering the milestone right on time would produce a line approaching the completion line horizontally. If we estimate an early milestone completion, the completion line would be approached with a downward trend. Although it is effective in predicting milestone progress, the chart is even more effective if used to develop actions required to eliminate any potential deviation from the baseline milestone schedule.

Constructing the Milestone Prediction Chart

Constructing a Milestone Prediction Chart begins with gathering the baseline schedule, current schedule performance, and all approved scope change This background information is needed to create the intent of the chart and to predict future milestone completion dates (see "Forgetting Trend Analysis?").

The chart is most effectively and efficiently constructed as a project team exercise, but before gathering the project team members, it is best to get a sense of the current status of each of the primary milestones and deliverables of the project.

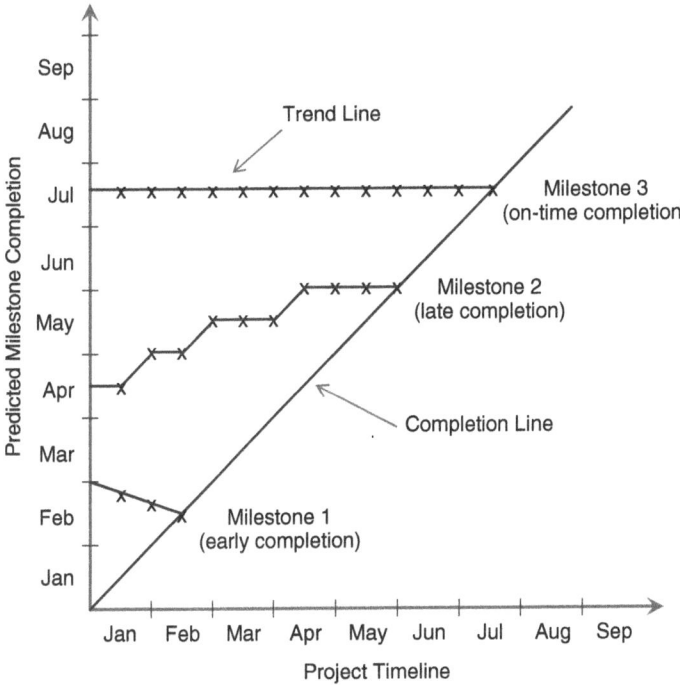

Figure 14.5: An Example Milestone Prediction Chart

Forgetting Trend Analysis: Déjà Vu?

Experience shows that the majority of projects place an emphasis on the evaluation of historical data to determine current schedule status as the core of their schedule management practices, leaving out the trend forecast to help predict *future* performance. As experts have been writing for decades, the primary purpose of project control is to prevent any sudden shocks to the project team, project sponsor, and key project stakeholders. Predictive analysis allows for corrective action before issues mount and recovery options evaporate. In other words, we predict trend because it is much better if we can predict ahead of time what is going to happen, "rather than just watch it happen." This makes trend analysis the single most important piece of information in project control.

Managers, especially those in industries where time-to-market is a competitive advantage, hate to suddenly hear that the schedule is going to slip. To them, it is much more meaningful to learn that ahead of time. Having a trend projection provides the managers with early warning signals so they can act while it is still possible to reverse unfavorable trends.

(Continued)

This doesn't mean the historical information is of little value. What we are saying is that history, of course, can't be influenced and the future can, so it is more important for the project team to use the historical information to forecast a future schedule trend and strategize actions to deliver the project as planned. It is for this reason that trend analysis should be the central piece of schedule management.

Gather Milestone Information

We suggest the project manager have a conversation with each milestone or deliverable owner to gain a good understanding of the current status at that point in time. As illustrated in Figure 14.5, not all milestones will be of immediate interest, as some will have already been achieved and some are too far in the future that gathering status is really meaningless. For the milestones in play, the project manager should inquire about the progress of each milestone or deliverable that the person is responsible for completing, whether the current progress differs from the plan, and if so, what causes and issues led to it, when they anticipate finishing the milestone or deliverable. If a variance to completion exists, inquire as to what the milestone owner thinks can be done to finish the milestones as planned.

Essentially, this initial conversation is a rehearsal for the project progress meeting that will focus on constructing or updating the Milestone Prediction Chart. This may look like a time-consuming activity for the project manager, but it is necessary to prepare the milestone owners for the progress meeting.

Periodic progress meetings instill the discipline and regularity in reviewing strides that are being made on the execution of a project. They should be called on a regular basis—once a month for a long project or once a week for a short project, for example. Each of the project milestone owners is required to attend the meeting, as well as other critical project team representatives such as quality assurance or a product manager.

Formally or informally, based on verbal or written information, milestone owners need to provide information on the status toward achieving the milestones that are of interest in the review. In this process, the understanding of the interfaces between milestones and the impact they may have on progress will be enhanced. Face-to-face, enriched exchange of information between milestone owners and others involved is the best way to get usable information.

Develop the Milestone Prediction Updates

During the initial milestone review meeting, the prediction chart has to be created. It can then be used in future review meetings. To begin, mark milestones from the baseline milestone schedule on the vertical axis of the chart; these are the "as planned" milestones. Next, draw the completion line. Because the vertical and horizontal scales use the same project schedule as the basis, the line has a 45° angle relative to both scales.

With the initial predictive chart created, the milestone review can commence, the owner of the first milestone crisply describes its actual progress, potential variance from the baseline and current issues causing the variance, and gives a preliminary prediction of the milestone completion. Relying on a schedule indicating the dependencies between milestones and activities, the team can see how the actual status of the milestone can impact the progress of other dependent milestones. A pointed discourse will normally unfold.

Owners of dependent milestones ask for more information and share their opinion about their actual progress, variance, and current issues. They also give a preliminary prediction of their milestone completion, analyze their actual impact on dependent milestones, and review possible future risks that may further affect the milestones.

At this point, preliminary predictions of milestone completions are made; however, the job is not yet complete. Steps must be taken to prevent possible slips. Everybody goes back to the drawing board and analyzes the dependencies between milestones. Corrective actions are formulated if needed and final predicted milestone completion dates are established. Those are marked on the Milestone Prediction Chart at the date of the prediction. A final analysis should be performed to determine if the project end date is affected by any slips in the milestone completion dates.

This exercise should be repeated every time schedule progress is reviewed. Once a milestone is completed, mark it on the project completion line. It should be noted that the Milestone Prediction Chart provides the project team a near-term view of milestone progress and is most effective when analyzing project progress within a one to three month window. Progress can only be gauged on milestones which are currently being worked on, or have dependencies on the milestones being worked on.

Using the Milestone Prediction Chart

The Milestone Prediction Chart is primarily designed to predict the completion date for major milestones. It was originally designed to focus on six to seven milestones at the highest level of the project, small or large.

The chart is best used in a rolling wave manner where detailed milestone progress is analyzed within a one to three month window, which repeatedly moves through time with the progression of the project.

To move beyond the near-term window, an analysis of the dependencies between the milestones has to be performed. Geared with such understanding, the project manager can use the chart to formulate control strategies and report project progress to stakeholders.

The Milestone Prediction Chart creates a sense of predictability in achieving major project events, or milestones. Through an environment of due to the disciplined progress reviews, the chart helps predict completion dates for major milestones on a regular basis, helping identify trends and leading to actions to correct possible negative trends. This proactive approach enables the project team to look into the schedule of milestones on the higher level of the project. Because of this both project managers and top-level managers find this too useful (see "The Tool of Choice").

The Tool of Choice

When a group of senior project managers from a leading sportswear manufacturer engaged in a presentation titled "A Few Good Schedule Management Tools," it took them all of about ten minutes to unanimously choose the tool they would prefer to use on their next project to manage their project timeline.

The overview presentation included details on the Slip Chart, Jogging Line, B–C–F Analysis, Earned Value Analysis, Milestone Analysis, and Milestone Prediction Chart, and when it ended, the project managers were asked to pick one tool that they believed provided the greatest utility for providing a summary level view of project schedule progress. The Milestone Predictive Chart was chosen as the tool of choice.

B–C–F ANALYSIS

The B–C–F Analysis, which stands for Baseline–Current–Future Analysis, compares the baseline project schedule with the current schedule performance and a schedule derived from the worst-case future scenario (see Figure 14.6). As a result, schedule performance trend is detected, or in other words, where we predict the schedule performance is headed. Most importantly, if the trend is unfavorable, it forces a project team to design actions to prevent it, which is the ultimate purpose of all proactive schedule management tools.

The B–C–F Analysis enables the project team to visualize the future of their project schedule and devise actions necessary to get there. The analysis is more applicable in smaller and medium-sized projects than in large and complex projects.

The B–C–F Analysis provides project managers with the ability to foresee the estimated worst-case scenario when schedule slippage occurs and risks begin to mount. For project managers who have been assigned the responsibility of turning a failing project around, there is no better schedule management tool available to help analyze the current situation and forecast future scenarios.

Performing the B–C–F Analysis

Performing a B–C–F Analysis begins with gathering the baseline schedule current schedule performance, and all approved scope changes. For the *baseline schedule* it is a good idea to work with a Jogging Line added to the baseline schedule (see previous section) so you can see how much each project task is ahead or behind the baseline schedule, what the current issues are, and what remains to be completed.

Prepare the Current Schedule

Using the baseline schedule with a Jogging Line the project manager and task owners can forecast the amount of change to the baseline schedule (if any). This will produce

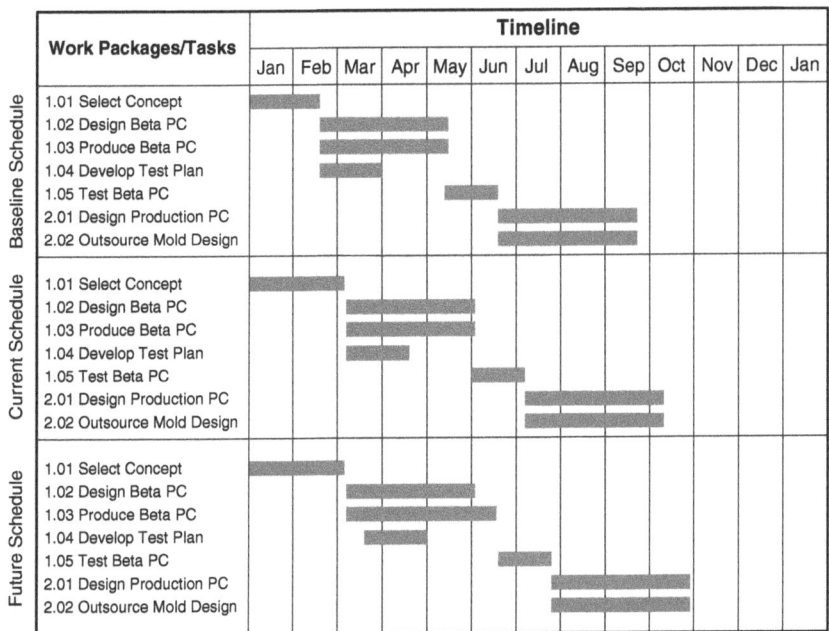

Figure 14.6: Example B–C–F Analysis

a new duration for each task, when it was started, and when it will be finished. Using this information, a new baseline schedule is computed. This is the *current schedule*, reflecting how the project is expected to unfold in the future given the task performance to date.

Develop a Future Schedule

To develop the future schedule, the project manager should ask each work package lead the following question: "What is the worst thing that can realistically happen based upon where we are now? This is a worst-case scenario planning exercise. Visualizing threats, dangers, and risks that the project may encounter helps in developing the future schedule. This exercise can include the risks on the risk register, but it should also include worst case scenarios.

Once the worst-case scenarios have been identified, the next step is to figure out how they can impact the baseline schedule. For this, the project team needs to look closely at the dependencies between tasks in the current schedule. Will the risks impact tasks on the critical path? If the impacted tasks are on a non-critical path, will the total float be consumed? How much could it push out the completion of impacted tasks? If the critical path is impacted, how much will the project completion be pushed out? When the team develops answers to these questions, take the current schedule and extend task durations accordingly to obtain the *future schedule*.

Looking into the Past

When developing the future schedule in the B–C–F Analysis, a project team can consider issues and risks encountered on previous projects. It is possible that an event that occurred in the past may affect current projects.

For each risk or issue the project team asks: "Could this happen on our project? Would its impact on our project be different from the impact it had on the past projects? Would actions taken on the past projects work on our project, or would we have to use different actions to defend against such events?"

Take Action

The primary purpose of the preceding steps is to equip the project manager with an early warning signal. A signal that says, "take action to resolve the issues causing the current schedule variance and mitigate risks to the future schedule." If the issues cannot be eliminated, there is a need to rechart the future schedule and find alternatives that would lead the project team to deliver as expected per the baseline schedule. One option is to try to fast-track the project.

To fast-track a project, perform the following activities and evaluate their effects on the overall project schedule:

1. Go back to the future schedule and focus on hard and soft dependencies between activities
2. Turn any of the sequential activities with hard dependencies into overlapping activities as much as possible, while still observing the hard dependencies. This may include changing a finish-to-start dependency into a finish-to-start with a lead, or into a finish-to-finish with a lag.
3. Reexamine all soft dependencies in order to overlap as many activities as much as possible. Given the soft dependencies, pick activities that can be performed out of the sequence established in the schedule.
4. Do a reality check on your schedule. Make sure that the fast-tracked schedule takes into consideration resource availability and that the new flow of work doesn't add new risks.

Although all of these changes can help accelerate the project, you need to search for more opportunities. Crashing the schedule by adding in more resources to reduce duration of activities on the critical path can provide additional time savings (see the "Schedule Crashing" section later in this chapter). Fast-tracking and crashing, however, may substantially increase the number of critical activities and paths, putting more pressure on project time management.

Using the B–C–F Analysis

The B–C–F Analysis is more applicable to smaller and medium-sized projects than to large and complex projects. For larger projects with many dependencies, the application of the B–C–F Analysis may be too cumbersome and time consuming.

A reasonably skilled and prepared project team of smaller size can prepare a B–C–F Analysis for a 25-activity schedule in 45–60 minutes. The necessary time will expand as the project size increases.

When using a B–C–F Analysis, the following guidelines should be used to ensure maximum effectiveness:

- Work with the work package owners to help them give accurate project schedule performance information
- Strive to develop good enough, not perfect, current and future schedules
- Insist on maximum interaction among task owners in progress meetings to help them understand how they impact each other and the project
- Observe which task owners tend to be too optimistic or pessimistic when forecasting their completion times. They may need personal coaching to overcome these tendencies.

Keep in mind that like the Jogging Line, the B–C–F Analysis provides a snapshot in time of project performance and must be performed on a periodic basis.

SCHEDULE CRASHING

Schedule Crashing is a method of shortening the total project duration without changing the project logic, which means that the sequence of dependencies between project activities remains the same. To compress the duration, the project usually deploys more resources in performing activities. As a consequence, the total project cost grows.

Performing Schedule Crashing

Schedule Crashing requires a process of disciplined and patient steps. We will demonstrate this process by showing how a schedule is crashed from seven to four days.

The following inputs are needed in order to effectively crash a schedule.

1. The baseline schedule is needed with dependencies and resources.
2. The current schedule performance, which when compared to the baseline identifies the amount the schedule needs to be crashed.
3. Any scope changes that have not already been incorporated into the baseline.
4. Resource loading and availability, along with the associated labor rates and other relevant cost information, understand the cost impact of the crashed schedule.

Develop a Normal, Cost-Loaded Schedule

This is the baseline schedule developed during project planning. Here resources are assigned to project activities, and their costs are calculated. Without this resource and cost information, Schedule Crashing the way we define it here is not possible. In Figure 14.7 we give an example of an original schedule (starting position) with duration and cost for each activity in the table. The darker tasks indicate the critical path.

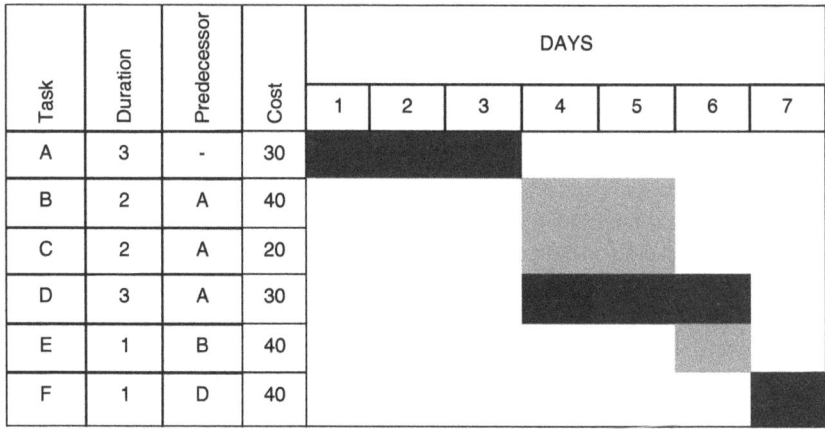

Task	Duration	Predecessor	Cost	DAYS						
				1	2	3	4	5	6	7
A	3	-	30							
B	2	A	40							
C	2	A	20							
D	3	A	30							
E	1	B	40							
F	1	D	40							

Figure 14.7: Schedule Crashing Example—Original Schedule

Identify the Cost and Duration to Crash the Schedule

While preserving the sequence of dependencies between project activities, perform the following steps:

1. Estimate the shortest possible time to complete each activity.
2. Ask the activity owners and team the following questions: "What resources do you need to complete each activity in this amount of time?" and "How much does it cost?" Also, it may take multiple iterations to develop these estimates, called crash durations. In the process, various challenges will be encountered. For example, some activities cannot be completed in shorter time than the normal schedule shows, or some activities will require additional human and non-human resources. Some resources that the project needs may not be available even though there is budget for them.
3. Compute the cost/time slope for each activity. Some activities are more costly to shorten than the others. Calculating the cost/time slope will show the cost of reducing the duration of each activity by one day. Use the following formula to compute the slope for each activity (see Table 14.1):

$$\frac{cost}{time\ slope} = \left(\frac{crash\ cost - normal\ cost}{normal\ time - cost\ time} \right)$$

This creates the basis to identify the most cost-effective activities to crash. Table 14.1 shows the duration and cost for each activity for our example.

Focus on the Critical Path Only

The critical path is the longest path in the network schedule, composed of activities whose float is zero. The duration of the critical path is the minimum time to complete all

Table 14.1: Crashed Schedule Duration and Cost Amounts

Activity	Duration (Days)		Cost ($)		Cost/Time Slope*
	Normal	Crash	Normal	Crash	
A	3	2	30	50	20
B	2	I	40	60	20
C	2	I	20	80	60
D	3	I	30	50	10
E	I	I	40	40	0
F	I	I	40	40	0
	Total: 7 days		Total: $200		

*Cost/Time Slope = (Crash cost − Normal cost)/(Normal duration − Crash duration).

project activities. Therefore, the duration of the critical path is equal to the total project duration. The only way to shorten the project duration, then, is to shorten duration of activities on the critical path. Simply, crashing the critical path duration by a certain number of days will translate into reducing the total project duration by the same number of days.

Crash the Most Cost-Effective Activities

When crashing, we want to do it with the minimum cost increase. For this reason, we don't choose to crash just any activity on the critical path. Rather, we focus on the most cost-effective one to shorten first, by selecting a critical activity with the least cost/time slope. In our example in Table 14.1 activity D has the least cost/time slope on the critical path, $10/day. Cut it by one day. As shown in Figure 14.8, Step 1, the schedule is now one day shorter, or six days long, and its cost equals the normal cost of the schedule plus the cost/time slope of the activity that was cut, which makes $210. In Figure 14.8 we show the critical path for each schedule as a bold line. Continue with cutting critical activities one day at a time, first those with the least cost/time slope—see Figure 14.8, Steps 2 and 3—until reaching the desired schedule duration of four days and its cost ($260).

Crash Multiple Critical Paths

As we crash activities on the original critical path (activities A–D–F in Figure 14.7), new critical paths appear. After cutting D in the first step, there are two critical paths—A–D–F and A–B–E. When this happens, to shorten the total project duration, shorten duration of all, in our case both, critical paths at the same time. This is why in Step 2, we cut activity A, which is on both critical paths, and in Step 3, we cut D on one critical path and B on another critical path.

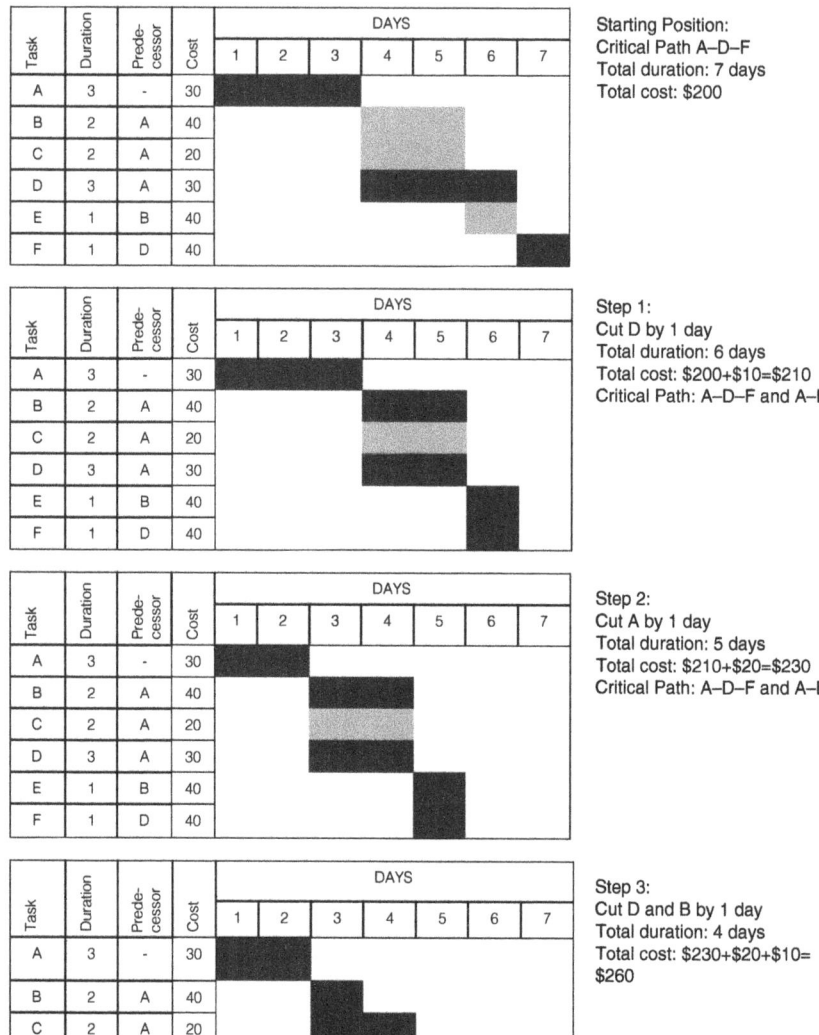

Figure 14.8: Schedule Crashing Example—Crashed Schedule

Failing to shorten one of them will leave the total project duration unchanged; simply, the longest path(s) determines the total duration. As multiple critical paths are crashed simultaneously, follow the rule of first crashing the least slope activity on each of the paths. Having multiple paths is why we enforce the rule of "crash one day at a time."

Using Schedule Crashing

Schedule Crashing is primarily a method involving two project scenarios. In the first, the project is in the planning stage, the execution has not started yet, and the project team proposes a schedule for management approval. Management finds the schedule too slow and demands it be shortened. To accomplish this, the team goes back to the drawing board, employing Schedule Crashing.

The second scenario occurs when the project is underway and the schedule slips. To catch up, the team may use the Schedule Crashing method. While for both scenarios the team can apply Schedule Crashing alone, many teams combine it with fast-tracking. Remember, fast-tracking changes the project logic, altering the dependencies between project activities.

Schedule Crashing provides a way to correct for negative schedule performance variances. Step by step, it shows which activities to crash, what resources it takes, and how much it costs. To all organizations cherishing time-to-market speed, or more generally, fast cycle times, this capability is a significant benefit (see "Five Golden Rules of Schedule Crashing").

Five Golden Rules of Schedule Crashing

1. Crash only activities on the critical path
2. First crash critical activities that are the least costly to crash (the least cost/time slope)
3. Crash by one time unit of the schedule at a time (one day at a time for example)
4. When there are multiple critical paths, crash all of them simultaneously
5. Don't crash non-critical activities.

CHOOSING YOUR SCHEDULE MANAGEMENT TOOLS

The tools presented in this chapter are designed for different project situations. Most project managers find it useful to employ one to three schedule management tools. To help in this effort, Table 14.2 lists various project situations and identifies which tools are geared for each situation. Consider this table as a starting point, and create your own custom project situation analysis and tools of choice to fit your particular project management style.

Table 14.2: A Summary Comparison of Schedule Control Tools

Situation	Slip Chart	Buffer Chart	Jogging Line	Milestone Prediction Chart	B–C–F Method	Schedule Crashing
Small and simple projects			✓	✓	✓	
Task-level progress reviews	✓	✓	✓	✓	✓	
Low-level detail needed	✓	✓	✓			✓
Short time to train how to use the tool	✓		✓	✓	✓	
Focus on highly important events				✓		
Large, complex, and cross-functional projects	✓	✓	✓	✓		✓
Fast projects		✓				✓
Projects of strategic importance		✓		✓		
Focus on top-priority activities	✓	✓		✓		✓
Summary detail needed	✓			✓	✓	
Display trend	✓	✓		✓	✓	
Provide predictive analysis		✓		✓	✓	
Little time available for schedule management			✓			
Analyze failing project	✓	✓			✓	
Correct project delays						

References

1. Kahn, K.B., Slotegraaf, R.J., Kay, S.E., and Uban, S. (ed.) (2012). *The PDMA Handbook of New Product Development*. Hoboken, NJ: John Wiley & Sons.
2. Goldratt, E.M. (2002). *Critical Chain*. Great Barrington, MA: North River Press.

15

COST MANAGEMENT

Project cost management involves management of the processes required to ensure that project work is accomplished within an approved budget, assessing changes to the budget based upon changes to the project, establishing a new budget baseline when necessary, and ensuring budget variances do not exceed the established threshold. Generally, there is minimal concern if work is completed under budget, so primary attention is focused on preventing budget overruns. The primary exception being if a budget underrun indicates work is not being performed as planned.

Budget variances are common and need to be a primary focus of project managers during the execution stage of a project (see "Common Reasons for Cost Overruns"). Cost variances caused by scope changes or major resource adjustments are fairly easy to detect. However, most budget under and overruns occur one day at a time and are more difficult to detect. Project managers need to be vigilant about preventing budget variances from accumulating to an unacceptable level. To do so, they need to have tools that provide early warning of potential and realized budget slips.

The tools presented in this chapter provide a variety of/options for early-warning detection of budget variances. Some tools are more suited for smaller projects, some for larger projects, others for complex projects, and still others for simple projects. Since the type of project a project manager will be called upon to manage will likely vary over the course of his or her career, it is recommended that each tool become a part of each project manager's PM Toolbox.

Common Reasons for Cost Overruns

It is not uncommon for projects to exceed their budget. No single reason explains all of the cost overruns on projects. However, there are a number of reasons that show up consistently in project cost management studies.

(continued)

Project Management Toolbox: Tools and Techniques for the Practicing Project Manager, Third Edition. Cynthia Snyder Dionisio, and Russ J. Martinelli.
© 2025 John Wiley & Sons, Inc. Published 2025 by John Wiley & Sons, Inc.

1. *Insufficient funds.* One of the main reasons for budget overruns is underfunding. This refers to the act of not allocating an adequate amount of budget to a project to begin with. Without adequate funding, project success becomes a matter of wishful thinking.

2. *Incorrect cost estimates.* Accurate cost estimation is crucial to effective cost management. If the cost of the project is under estimated during the planning stage, a budget overrun will eventually occur unless the scope of the project is reduced to match the estimated cost. This is especially true in environments where top managers of a firm set cost targets or not-to-exceed thresholds that artificially limit the amount of budget that is estimated. Likewise, if the cost of a project is over estimated, a budget underrun will occur.

3. *Scope increases.* Increases in scope on projects frequently cause cost overruns. These changes map directly to project requirements. Normally scope increases are a result of missed requirements during project definition and planning or new requirements that are introduced during project execution. New requirements demand additional work and additional work demands additional project cost.

4. *Extended project schedules.* If project activities take longer to complete than originally planned, the project schedule usually has to be extended to reflect the correct amount of time required. Schedule extensions mean additional costs caused by additional unplanned resource hours. This is especially true as the complexity of a project increases. The higher the project complexity, the higher the number of interdependencies between team members and activities, therefore the higher the probability that interconnected tasks and activities will take longer than planned.

5. *Lack of risk management contingency.* Failure to establish a contingency fund for events that are likely to occur or with severe impact on project completion accounts for many cost overruns. Failure to plan for and conduct risk management creates the assumption that the best-case execution outcome will occur, and no unexpected impacts to the project will occur.

6. *Poor cost management.* Failure to effectively manage the cost aspects of a project have resulted in many project cost overruns. They can include the failure to clearly identify who is responsible for managing project cost, who has the authority to approve changes to the project budget, how cost performance will be measured and reported, and the identification of triggers to signal the need for corrective actions.

Even though changes to the project budget are common, they are not a foregone conclusion if effective project cost management practices are followed. Cost management practices involve cost performance measurement, forecasted budget at project completion, change impact analysis, recommended corrective actions, and making updates to the project management plan and budget baseline.

Beginning with the Budget Consumption Chart, the following tools are designed to assist project managers perform their project cost management practices effectively.

BUDGET CONSUMPTION CHART

The Budget Consumption Chart is a graphical representation of the project expenditures as they occur over the project timeline. The chart is a very effective tool for project managers to communicate current budget consumption status.

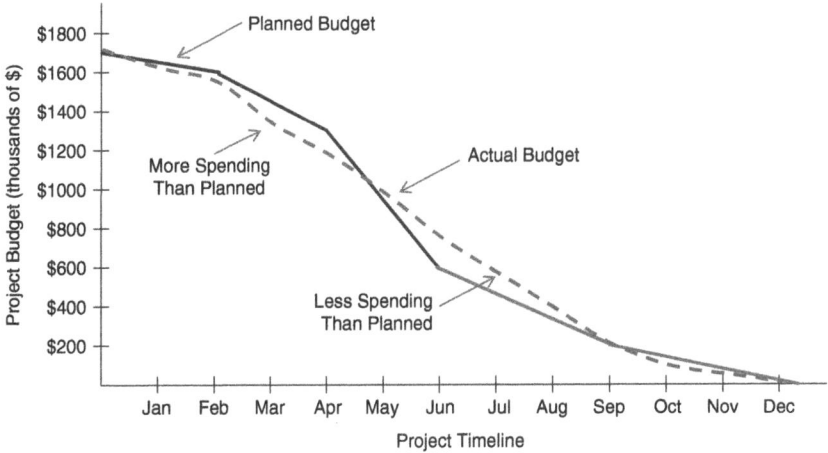

Figure 15.1: Example Budget Consumption Chart

As illustrated in Figure 15.1, Budget Consumption Charts are represented as standard *x–y* charts, with the *x*-axis representing the project timeline and the *y*-axis representing the project budget.

Project managers should keep in mind that the Budget Consumption Chart is not designed as a comprehensive tool to measure overall project cost performance, especially as it directly relates to schedule performance. Rather, it is only focused on measurement of project budget spending and serves as a good tool for preparing the project manager for a broader analysis of project cost management status.

Developing the Budget Consumption Chart

Developing the Budget Consumption Chart requires a detailed understanding of the project scope, the resources needed to complete the work, the cost of the resources, and the overall budget for a project. This information is normally integrated into a cost baseline (see Chapter 8). Additionally, the incremental cost of any approved scope changes that have not been incorporated into the budget baseline will need to be evaluated for impact to the project budget and incorporated into the chart.

Create the Budget Consumption Template

The Budget Consumption Chart simply represents the amount of project budget available to the project manager at any point in time. Begin by laying the project timeline across the *x*-axis of the chart. If a rolling-wave or iterative project execution methodology is being employed, only show the portion of the project timeline which is pertinent. Time zero on the *x*-axis represents the point in time when project execution begins, or the point in time when a particular execution iteration begins.

The *y*-axis represents the project budget and should be distributed across the axis in equal amounts. The total project budget is represented at the highest point of the *y*-axis with zero dollars represented at the intersection of the *y*-axis and the *x*-axis. The budget should be represented in this manner in order to illustrate budget consumption over time, with full budget available at project execution start, and zero budget available at planned project closure.

Draw the Planned Budget Consumption Line

Using the information contained in a project scheduling tool (the Gantt Chart is recommended), plot the planned budget expenditure over the project timeline. Next, connect the periodic budget expenditure or value points with a contiguous line that represents the planned amount of budget remaining at the normal project review intervals, or major deliverable dates. At project closure, the planned budget should be zero. Note that, as illustrated in Figure 15.1, this is normally not a completely linear representation due to the variations of resources, materials, and other direct costs; the exception being projects that have constant level of effort resource loading. The chart should normally look like an inverted "s-chart."

Using the Budget Consumption Chart

The Budget Consumption Chart is used for both the planning and execution stages of a project. First, the chart represents the amount of project budget available over a specified period of time. During project planning, the chart can be used as an early indication of project spending under various scope, resource, and timeline scenarios. This is especially useful in situations where budget is constrained, or available at various points in time—such as projects that rely on government funding from multiple sources or multiple fiscal years. Second, the chart provides a visual representation of the amount of budget spent and amount of budget still available at any point on the project timeline.

At a glance, the chart can indicate whether a project manager is spending more or less budget than planned. Ideally, you would like the actual line to lie extremely close to the planned line. However, we do not live in an ideal world and project spending seldom occurs as planned. Referring to Figure 15.1, we see a more realistic scenario. When the actual line moves above the planned line, it is an indication that the team is spending less than was planned, and therefore more budget is available than planned. Conversely, when it is below the ideal one, more project budget is being consumed than planned.

When variations above and below the planned line show up, it may indicate the need for corrective actions. However, how do you know when a corrective action is needed? Good cost management practices involve the use of variance thresholds that operate as corrective action triggers. One such trigger might be percent above or below planned budget consumption. For example, if current budget consumption is 10% more or 10% less than planned at any point in which the project status is reviewed, it would indicate that corrective action is needed.

You should restrain yourself from reacting too quickly. Keep in mind that you are only looking at the project from a single perspective (budget consumption), and that you are missing the project schedule, resource, and risk perspectives. Recall that earlier we stated that the Budget Consumption Chart is best used to prepare the project manager for a broader discussion on project performance. Because it only provides a single

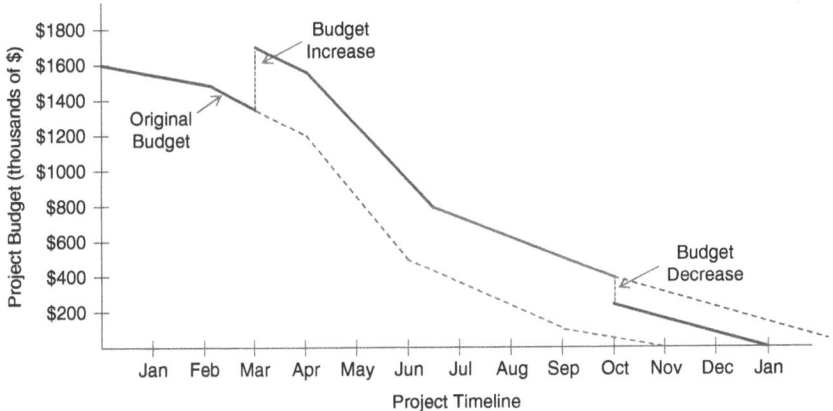

Figure 15.2: Budget Consumption Chart with Funding Changes

perspective of performance and taking corrective action on a single perspective is a risky proposition. Instead, we recommend that the Budget Consumption Chart be used in tandem with other cost management tools presented in this chapter that provide a more integrated view of project performance.

Variations

The project scenario represented in Figure 15.1 contains a major assumption. It assumes that the entire project budget is available when project execution begins. Many times this is not the case, such as in the world of government procurement projects. Figure 15.2 illustrates how the Budget Consumption Chart can be modified to include changes in project funding.

When an increase in project funding occurs, the project budget increases and the planned budget line on the chart is increased vertically by the amount of incremental budget received. In contrast, when project funding is lost, perhaps due to de-scope of work, the planned budget line is decreased vertically by the amount of budget no longer available to the project manager. The reasons for the budget fluctuation should be recorded. Once the budget baseline is reset, normal use of the Budget Consumption Chart resumes.

EARNED VALUE ANALYSIS

The best method for determining the overall project performance status for large projects generally, and cost status specifically, is Earned Value Analysis (EVA). EVA is a project performance measurement technique that integrates scope, time, and cost data. It periodically records the past performance of a project and forecasts its future performance (see Figure 15.3).

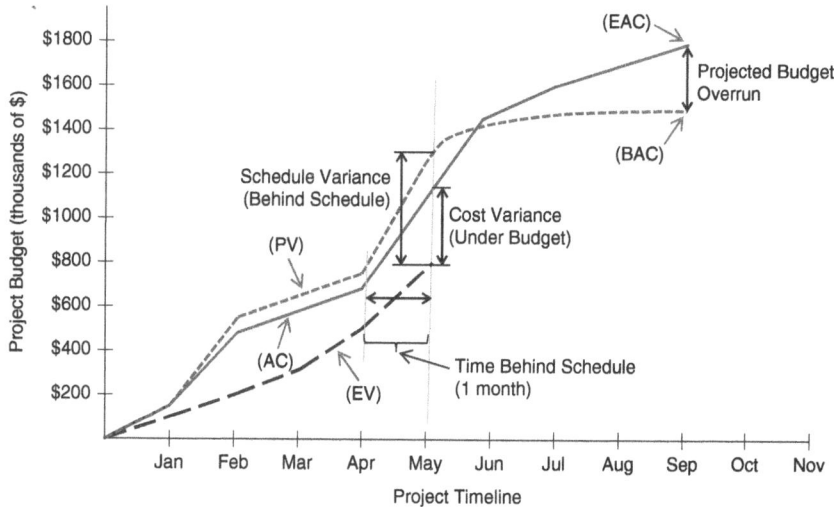

Figure 15.3: An Example Earned Value Analysis Chart

As the project progresses, EVA measures a project's schedule and cost performance to determine whether the project is ahead or behind plan (schedule and cost variances) and why. Then, final project costs (estimate at completion) and completion date (schedule at completion) are predicted based on current performance.

While the practical elegance of such an approach comes from EVA's integration of project scope, cost, and time, its special value is in providing project managers with the early-warning signal about possible problems in the future and giving them an opportunity to devise and take needed corrective actions while there is still time to fix the problems.

In summary, EVA strives to establish the accurate measurement of physical performance against a plan to enable the reliable forecast of final project costs and completion date.[1] We describe the sequence of steps to explain the conceptual simplicity of performing an Earned Value Analysis. To begin, refer to Table 15.1 which defines the basic Earned Value Analysis terminology and Table 15.2 which defines the key formulas.

Performing Earned Value Analysis

EVA is primarily used in large government projects and in the construction industry. Private business generally performs work through small and medium projects, which can use EVA in a simpler form. Fully valuing both approaches, for large government and smaller private projects, we will first focus on a comprehensive, but simplified approach using current terminology.

Performing an Earned Value Analysis needs to be precipitated by good project management practices, such as fully defining the project scope with a work breakdown structure (see Chapter 6), a baseline schedule using the critical path method (see Chapter 7), and a time-phase project budget, AKA cost baseline (see Chapter 8).

Table 15.1: Fundamentals of Major Earned Value Measurement Methods

Term	Acronym	Description
Planned value	PV	The estimated cost for a work package or control account
Actual cost	AC	The cost incurred for the work completed on the report date
Earned value	EV	The value of the work performed for a work package or control account as of the report date
Cost variance	CV	The difference between the value of the work performed (EV) and the costs incurred (AC)
Schedule variance	SV	The difference between the amount budgeted for the work completed and the value of the work completed
Cost performance index	CPI	An index that reflects the value of the work accomplished (EV) compared to the cost of the work accomplished (AC)
Schedule performance index	SPI	An index that reflects the value of the work completed (EV) compared to the estimated cost of the work completed (PV)
Budget at completion	BAC	The original estimate of the total cost to complete a work package, control account, or the project
Estimate at completion	EAC	The current estimate of the total cost to complete a work package, control account, or the project
Estimate to complete	ETC	The current estimate of the amount of funds required to complete the remainder of the project
Report date		The Data Date or point in time at which the Earned Value Analysis is performed

Table 15.2: Key Earned Value Analysis Formulas

Term	Formula
Cost variance	$CV = EV - AC$
Schedule variance	$SV = EV - PV$
Cost performance index	$CPI = EV/AC$
Schedule performance index	$SPI = EV/PV$
Estimate at complete*	$EAC = (BAC - EV) + AC$ $EAC = BAC/CPI$
Estimate to complete	$ETC = EAC - AC$

*There are multiple ways to calculate the EAC, these are two of the quickest ways. EVA professionals will choose a method that best reflects the expected future performance.

The WBS provides the basis for scope of work. Each work package will be carefully analyzed to determine when the work will be performed. Details about beginning and ending points of work, as well as their durations, will be documented in the schedule. Such information, along with the approved budgets for work, provides the *planned value*. As the project implementation unfolds, physically completed work is evaluated and the *earned value* determined. Both the *planned value* and *earned value* are derived from the project schedule information and are critical for successful EVA.

Set Up a Performance Measurement Baseline

A performance measurement baseline (PMB) is established to determine how much of the planned work the project team has accomplished at any point in time. Establishing a PMB involves three tasks: (1) determining points of management control and who is responsible for them, (2) selecting a method for measurement of earned value, and (3) establishing the baseline.

The foundation for the performance measurement baseline is the WBS with fully defined project scope, allocated resources, and a project schedule for performance. Given that the WBS has elements on multiple levels, you have to decide which elements (on which level) will be management control points. These points are called *control accounts* (CA). Although at first sight this might seem like a confusing term, in actuality its concept is simple—a CA is a point at which we measure and monitor performance. The makeup of a CA is defined in the example titled "Key Components of a Control Account."

Key Components of a Control Account

Narrative scope definition
Location in WBS (e.g., level 1, 2, or 3 in a WBS with level 0 for the project)
Constituent work elements (e.g., level 2, 3, or 4 for work packages)
Timeline (e.g., begin/end dates of each work package)
Budget (resource hours, dollars, or units for each work package)
Owner: the person responsible for the CA (e.g., Software PM)
Type of effort (e.g., nonrecurring or recurring)
Methods to measure EVA performance (e.g., weighted milestones)

CAs may be located on a selected level of a WBS—at level 1, 2, or 3 (the project is level 0 of the WBS), or all the way down to whatever is chosen as a lowest level to exercise management control. The essence is that a CA is a homogeneous grouping of work elements that is manageable. A CA has a clearly defined narrative scope, location in the WBS, constituent work elements, timeline, and budget. Although the budgets are often expressed in dollars, they can be expressed in resource hours as well. Because so many project managers manage only resource hour budgets, we will use hours in our examples. To ensure accountability for the budgets, each CA should be assigned to a person responsible for its performance.

Figure 15.4 illustrates how a WBS includes control accounts (CA) at Level 1, followed by work packages (WP) that represent deliverables beneath the control accounts. Each control account has a Control Account Plan (CAP) and an accountable party who is

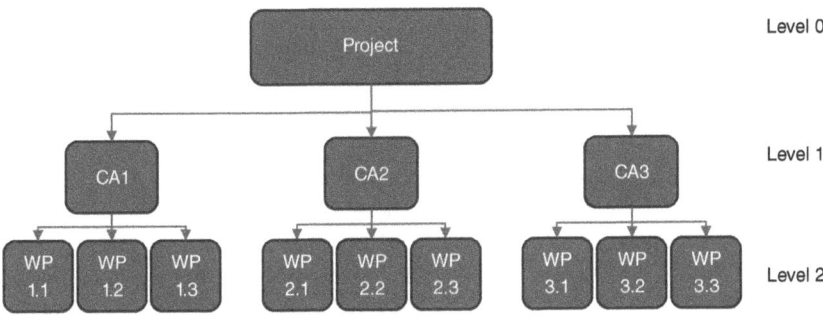

Figure 15.4: Control Accounts and Work Packages

assigned to the control account. This person is referred to as a Control Account Manager (CAM).

Measuring a CAP's performance calls for well-defined methods of measurement. While we review several such methods (see Table 15.3), hard-and-fast rules for selecting the appropriate one do not exist. The project manager, the project team, and CAP managers should focus on the ease and accuracy of measurements that can be consistently applied to appropriately support their specific project needs.

Table 15.3: Fundamentals of Major Earned Value Measurement Methods			
Type of Method	**When to Use**	**Major Advantage**	**Major Disadvantage**
% complete	Well-defined work packages; management reviews in place; nonrecurring tasks	The easiest method to administer	Made purely on a subjective basis
Fixed formula	Work packages are detailed and short-span; nonrecurring tasks	Easy to understand	Rather subjective
Weighted milestones	Work packages run two or more performance periods; nonrecurring tasks	Perhaps the most objective method	Difficult to plan and administer
% complete with milestone gates	Works in any industry, on any type of project; nonrecurring tasks	Both easy and objective	Requires time and energy to define meaningful milestones
Earned standards	Preestablished standards of performance; nonrecurring or recurring tasks	Perhaps most sophisticated of all methods	Requires the most discipline
Equivalent units	Long performance periods; nonrecurring or recurring tasks	Simple and effective	Requires a detailed bottom-up estimate

The *percent complete* method uses a periodic (monthly or weekly) estimate of the percentage of completion of a work package, expressed as a cumulative value (e.g., 65%) against the full 100% value of the work package. Hailed as a simple and fast method, which perhaps explains its wide popularity, the method has also been viewed as being overly subjective. Defining work packages' scope well and checking on accuracy of the estimates helps make the subjectivity reasonable.

Fixed formula by work package includes various options: 25/75, 50/50, 75/25, and so on. For example, a 25/75 formula means that when a work package is started, 25% of the package's budget is earned, while the completion of the package earns another 75%. Any combination that adds up to 100% is possible. This is a quick way of estimating, applicable in situations where work packages are short span and performed in a cascade type of time frame.

Weighted milestones divide a long-span work package into several milestones, each one assigned a specific budgeted value, which is earned when the milestone is accomplished. As objective as it is, the method's success hinges heavily on the ability to define meaningful milestones that are clearly tangible, budgeted, and scheduled.

Percent complete with milestone gates strives to balance the ease of percent complete estimates with the accuracy of tangible milestones. A work package of, say, 600 hours is broken down into three sequential milestones, each budgeted at 200 hours and placed as a performance gate. You are allowed to estimate the first milestone's earned value by percent complete up to 200 hours. To go beyond the point of 200, you need to meet predefined completion criteria for the first milestone. This procedure is repeated for subsequent milestones.

Earned standards is a method often applied by industrial engineers to establish planned standards for performance of work packages, which are then used as the basis for budgeting the packages and subsequently measuring their earned value. For example, the planned standard for producing a cup of lemonade at $0.20/cup is used to budget the work package including the production of 1,000 cups for $200. When 500 cups are produced, regardless of the actual cost the earned value is 500 cups × $0.20/cup = $100. Widely applied in repetitive types of project work, the method's foundations are the planned standards developed from historical cost data, time, and motion studies.

In *equivalent completed units*, a planned work package is earned when it is fully completed. Similarly, a planned portion of it is earned when completed. For example, a work package to build 5 miles (five units) of freeway is estimated at $3M/mile for a total of $15M. It is fully earned when all five miles are finished. Also, the completion of half a mile will earn $1.5M.

After this short review of the six methods, two things need to be mentioned in closing comments about measuring EVA performance. First, note that the work package is the place where the measurement is taken, while measurement for a CA is a summation of work packages' measurements. Second, there is no single best way to measure earned value for any type of project task. This means that different types of tasks will use different methods. For an example of a project where multiple methods are used, see Figure 15.5a. A design project consists of three CAs, essentially

a) Data on multiple methods of earned value measurement for a project

CAP	EV Method	Measure		Jan	Feb	Mar	Apr	May
Conceptual Design	% Complete Estimate	Planned		45	55			
		Earned		20	30	50		
		Actual		35	45	50		
Detailed Design	Weighted Milestones	Planned			100	100	50	
		Earned			100			
		Actual			115			
Prototype	Earned Standards	Planned				25	100	50
		Earned				25		
		Actual						
Total Project	Planned	Inc.		45	155	100	150	50
		Cum.		45	200	300	450	500
	Earned	Inc.		20	130	75		
		Cum.		20	150	225		
	Actual	Inc.		35	160	75		
		Cum.		35	195	270		

b) Cumulative performance curve for the planned value

c) Cumulative performance curves for the planned value and actual cost

d) Cumulative performance curves for the planned value, actual cost, and earned value

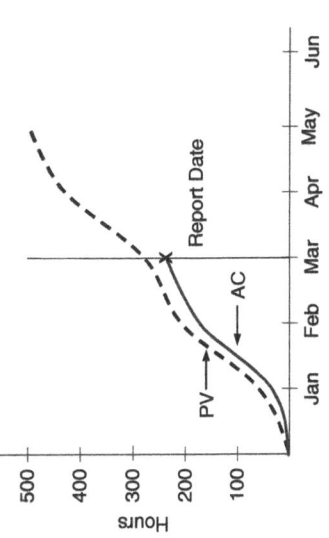

Figure 15.5: Performing Earned Value Analysis

three phases on level 1 of the WBS. Each of them applies a different method of EVA measurement—percent-complete, weighted milestones, and earned standards. With the EVA measurement method explained, it is time to establish the performance measurement baseline (PMB).

The PMB is a time-phased sum of CAPs. What is included in the CAPs depends on how companies define cost management responsibilities of their project managers. Many companies allow their managers of internal projects to manage only direct labor hours, which is our focus here. In that case, their CAPs and PMB will include only direct labor hours. On the other end of the spectrum are project managers whose job is to manage all project costs, as well as management reserves and profit. Accordingly, their PMB will reflect this situation.

In projects with a good degree of certainty, a firm PMB with detailed CAPs can be established before project execution begins. What if you have to start executing an uncertain project in which front-end CAPs are detailed out while the later ones cannot be planned for the lack of information? What if a CAP's scope starts changing? The answer for the first issue is the rolling wave approach; as you progress in executing the available detailed CAPs, you will generate more information that enables you to plan other CAPs. As for the second issue of scope changes, you should establish a PMB change control process. By carefully handling all changes to the scope, you will be able to update and maintain the approved PMB, a prerequisite to successful EVA.

For practical purposes of EVA, the time-phased PMB can be displayed as a cumulative performance curve representing the planned value over the project schedule. That is the curve shown in Figure 10.5b, developed for a design project whose performance data is given in Figure 10.5a.

Evaluate Project Results

This step compares the actual results of project performance with the plan (PMB). While the very measurement of performance occurs within individual CAs, you may monitor and periodically (e.g., weekly or monthly) evaluate performance results at three levels: within individual CAs, at some intermediary summary level (a WBS element beyond the CA), and at the project level. This step includes the following:

- Focus on schedule: evaluate schedule variance (SV) and schedule performance index (SPI).
- Focus on cost: evaluate cost variance (CV) and cost performance index (CPI).
- Identify the cause of the variances, if any.

Our example in Figure 10.5c shows the comparison, the cumulative performance curve for actual values against the performance curve for planned values. At the end of March in our example, the difference between the two curves, called the *spending variance*, only reflects whether the project stays within the approved budgeted hours.

It does not in any way determine the project's true cost performance status. If used for establishing the project's true cost performance status in Figure 10.5c, the comparison would mislead us by indicating that the actual cost performance is under budget (300 hours − 270 hours = 30 hours), a positive development. This couldn't be further from the truth, the project is in cost trouble, as we will soon see, and that can't be discerned using this planned value-versus-actual cost comparison. The reason for this false finding is that it does not incorporate the work that was performed (the EV). Thus Figure 10.5d is used.

A comparison of the earned and planned value at the end of March indicates the following:

$$\text{Schedule Variance}(\text{SV}) = \text{EV} - \text{PV}$$

$$\text{Schedule Variance}(\text{SV}) = 225 \text{ hours} - 300 \text{ hours}$$

$$\text{Schedule Variance}(\text{SV}) = -75 \text{ hours}$$

This negative SV means that the project has fallen behind its planned work. A look back at Figure 10.5d reveals two manifestations of the same SVs—one drawn vertically is expressed in budget units, the other horizontal in time units. Not surprisingly, you may prefer the one expressed in time units (days, weeks, months). It is generally easier to identify time units of delay by means of the schedule performance index. Before we get there, it is worth mentioning that any time SV is negative, the project is behind the planned work, and any time SV is positive, the project is ahead of the planned work. However, to get a true assessment of schedule performance, you must take into consideration the performance on the critical path.

Another task in evaluating the schedule position is calculating SPI. SPI quantifies how much actual earned value was accomplished against the originally planned value. In other words, it represents how much of the originally scheduled work has been accomplished at a certain point of time. An SPI equal to 1 means perfect schedule performance to its plan. Any SPI greater than 1 implies an ahead-of-schedule position to the original plan of work. SPI running below 1 reflects a behind-schedule position to the originally scheduled work.

At the end of March in Figure 10.5d, the situation is as follows:

$$\text{SPI} = \frac{\text{EV}}{\text{PV}}$$

$$\text{SPI} = \frac{225}{300}$$

$$\text{SPI} = 0.75$$

The SPI of 0.75 indicates that 75% of the originally planned work is accomplished. This means our project is behind the baseline plan of work. Since our reporting date at the end of March is the 90th day of the project, we can tell that our project is 22.5 days (25% of 90 days) behind the original work planned.

Schedule analysis in EVA deserves a word of caution. Specifically, anytime you find a schedule delay condition that includes negative SV and SPI that is less than 1, you should know that EVA schedule variance is not based on the critical path information and may be deceptive. Poor schedule performance of some work packages or tasks may be balanced by schedule performance of other work packages or tasks. Therefore, use your critical path schedule and risk analysis in conjunction with EVA schedule analysis. If the late work packages/tasks are on the critical path or are highly risky to the project, complete the work packages/tasks at the earliest possible date.[1]

Now we can move to our second area of interest in this step, calculate cost variance (CV) and cost performance index (CPI). At the end of March in Figure 10.5d, CV is as follows:

$$\text{Cost variance}(CV) = EV - AC$$

$$(CV) = 225 \text{ hours} - 275 \text{ hours}$$

$$(CV) = -45 \text{ hours}$$

and

$$CPI = \frac{EV}{AC}$$

$$CV = \frac{225}{270}$$

$$CV = 0.83$$

The purpose of CV is to indicate the differential between the earned value for the physically accomplished work and the actual cost to accomplish the work. Therefore, the positive CV means that the project is running under budget, while negative CV signals the project is spending more than planned, overrunning the budget. In essence, we are experiencing the latter, consuming 45 more hours than we have allocated for the amount of accomplished work.

CPI is a cost efficiency factor. By relating the physically accomplished work to the actual cost to accomplish the work, CPI establishes the cumulative cost performance position. When CPI is equal to 1, it indicates perfect cost performance to the original budget. Values of CPI exceeding 1 indicate under original budget position, while those less than 1 indicate over original budget position. In our example, the CPI reading of 0.83 tells us that the earned value for the physically accomplished work is only 83% of the actual cost to accomplish the work. Putting it differently, for every dollar spent, only 83 cents of value was accomplished.

Both CPI and SPI cumulative curves enable a very effective tracking of a project, as illustrated in Figure 15.6. Note that both rate and trend of the indices are crucial here. The key is using the cumulative data, rather than incremental data (weekly or monthly). Unlike the incremental data, which is prone to fluctuations, the cumulative data tends to smooth out the fluctuations and is very effective in forecasting the final project results, the focus of our next step.

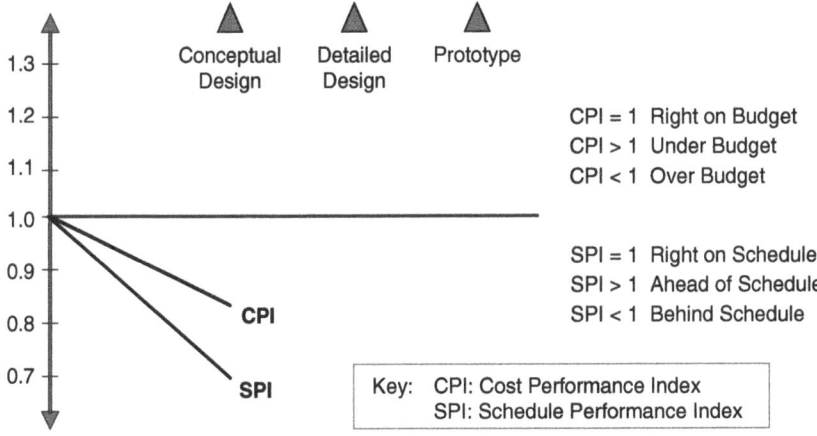

Figure 15.6: Tracking Cumulative Schedule Performance Index (SPI) and Cost Performance Index (CPI)

There is some controversy in using EVA to forecast schedule performance. Many project managers prefer a detailed analysis of the schedule with an emphasis on the critical path and the risk register. However, you can compute a quick forecast of the Estimate at Completion for time (EAC^t) using the information available for our example in Figure 10.5d at the end of March is as follows:

$$\text{Estimate at Completion}\left(EAC^t\right) = \frac{\text{original schedule}}{SPI}$$

$$EAC^t = \frac{150 \text{ days}}{0.75}$$

$$EAC^t = 200 \text{ days}$$

This quick method may be risky. As mentioned earlier, an EVA schedule delay condition that like in our example has a negative schedule variance and SPI that is less than 1 is not based on the critical path information and may be deceptive. Therefore, a better solution is to predict the completion date based on results of the critical path analysis in combination with reviewing the risk register.

Out of 20+ available formulas to estimate a project's cost at completion of the project, we will only look at two that are frequently used. An easy way to calculate the EAC is by dividing the BAC by the CPI as shown below:

$$\text{Estimate at Completion}\left(EAC\right) = \frac{\text{Budget at Completion}}{CPI}$$

$$EAC = \frac{500 \text{ hours}}{0.83}$$

$$EAC = 602 \text{ hours}$$

This means that at the end of the project we would need 602 hours to get this project done, 102 hours variance at completion (VAC) over the original budget. Clearly, this method relies on cost overrun to date and projects it to the end of the project.

A more rigorous method, based on our forecast on both the cost overrun and schedule slippage to date, is based on both the CPI and SPI:

$$\text{Estimate at Completion(EAC)} = \frac{\text{Budget at Completion}}{\text{SPI} \times \text{CPI}}$$

$$\text{EAC} = \frac{500 \text{ hours}}{0.62}$$

$$\text{EAC} = 806 \text{ hours}$$

With this method, we get an estimate at completion of 806 hours. Some researchers found that the low-end forecast is a reliable measure of the "minimum" hours, while the high-end method produces a forecast of the "maximum" hours we may need. Their claim that the high-end method is the most appropriate forecasting method should be contrasted with a recent study's finding that the low-end method is the most accurate. With such differing views using both methods to develop a range of final cost projections, in our case between 602 and 806 hours, makes sense. This is the absolute essence of prediction—produce a sanity check of the trend and final direction of the project (for major factors impacting the final results, see "Three Factors Influencing the Final Project Results").

In our example, the prediction is not good; actually, it is very bad, but its ultimate value purely depends on the willingness of management to act or not to act. If the option is to not act, the EVA is meaningless—it has no value whatsoever. Choosing to act by developing corrective actions rooted in the root causes of the problems is what EVA is designed for.

Using Earned Value Analysis

If there is one single, most compelling reason to use EVA, it is for its predictive ability, the ability to reasonably forecast the final project results during project execution, most of the time. We say most of the time because it is only somewhere at the 15% completion point and beyond that a sound, statistically reliable forecast becomes feasible for the project's cost at completion.

EVA is an option for any project, regardless of the industry and size. With the amount of resources at stake in large projects, a full-scale EVA can be easily justified. Simplified versions of EVA such as cost and achievement analysis (see the Variations section) are a good fit for smaller projects. In either case, a good measure of customization is recommended.

Three Factors Influencing the Final Project Results

1. *Sound project baseline*. Only when the scope is well defined, the schedule is realistic, and the budget is accurate can we expect a realistic forecast of the final project results.
2. *Actual status of the project*. The actual status of the project, as quantified by SPI and CPI, will be a vital factor in determining what final results the project will end up. Better SPI and CPI rates and trends indicate better final results.
3. *Corrective actions*. What will management do if the forecast is poor? Not believe it and do nothing? Or believe it and aggressively pursue corrective actions to alter the forecast? This is the moment of truth for management that will critically influence the final results.

The disciplined use of EVA can answer the question "Is the project on schedule, behind, or ahead of schedule?" Using the schedule variance and SPI in conjunction with the critical path method can reliably answer this question. In a similar manner, the cost variance and CPI play the crucial role in establishing the true cost position of the project by finding whether it is on, over, or under budget.

In particular, research of the past use of EVA indicates that cumulative cost performance index for larger projects becomes very stable at the 15% completion point in the project.2 Simply, this means that early in the project the cost performance index exhibits a consistent pattern, enabling reliable forecasts of the project cost at completion. Similarly, the schedule performance index combined with the Critical Path Method can be used for predicting the final completion date. Hence, you can periodically ask, "Given my current performance, what will be my final costs and completion date?" The answer offers trend performance and, if the trend differs from the baseline plan, an early-warning signal.

Variations

There are several cost control tools that are a simplified version of EVA. Two such tools that enjoy a high level of popularity are Milestone Analysis and Cost and Achievement Analysis.3 Overall, their appeal is in that they use simple terminology and straightforward process, which is perhaps why they are so time-efficient. Because of a perception in the PM community that the Milestone Cost Analysis is a tool of its own, it is described as a separate tool in the next section in this chapter. The Cost and Achievement Analysis is briefly covered in the following example is illustrated in Figure 15.7.

COST AND ACHIEVEMENT ANALYSIS

Project Name: ——————————   Estimate Date: ——————————

1	2	3	4	5	6	7
Task No.	Task Description	Budget (hours)	Percent Complete to Date	Achieved to Date (3) X (4)	Actual Cost to Date	Predicted Final Cost (hours) $(6) + \frac{(3) - (5)}{(5) + (6)}$
12	Prepare Bill of Materials and Routing	8.0	40%	3.2	5.0	12.5
	Total	312.0	36.4%	113.6	118.0	206.1

Figure 15.7: Example Cost and Achievement Analysis

Based on the scope and schedule for a task, its budget (same as the planned value in EVA) of resource hours is defined. Multiplying the budget by the percent complete will produce the achieved value (equivalent to the earned value in EVA). Actual consumed hours (equivalent to actual cost in EVA) to complete the scope defined by the achieved value are recorded as well. Values for the budget, achieved, and actual cost are then used to predict the final cost for a task. Doing this on a regular basis for each task, in cumulative terms, allows you to sum them to produce budget, achieved, and actual values for the whole project and predict the project's final cost. The approach offers a great way to be proactive in smaller projects.

MILESTONE COST ANALYSIS

Milestone Cost Analysis compares the planned and actual cost performance for milestones to establish cost and schedule variances as measures of the project's progress (see Figure 10.8). A Milestone Cost Analysis is a simplification of EVA. A milestone's cost is planned and tracked on the y-axis, and its schedule on the x-axis. The gap between the milestone's planned and actual cost provides the cost variance. Similarly, the schedule variance is obtained through the differential between the planned and actual schedule for the milestone. Both the planned and actual values are portrayed by cumulative curves. These two curves—as opposed to EVA's three curves of plan, actual, and earned values—are made possible by using milestones as a platform for

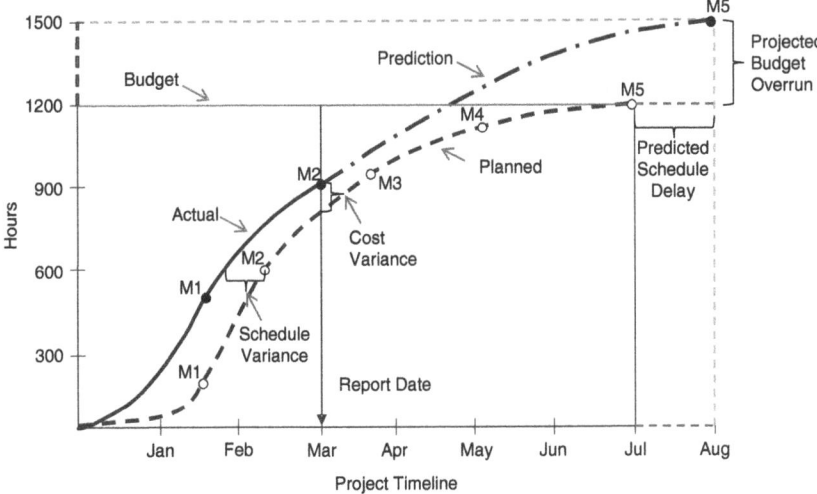

Figure 15.8: An example Milestone Cost Analysis

the integration of scope, schedule, and budget. Although effective in tracking project progress, Milestone Analysis is far more effective when used predictively to estimate the final project cost and completion date.

Performing a Milestone Cost Analysis

The quality of the Milestone Cost Analysis is dependent on the quality of the input. Therefore, a fully defined project scope, and detailed project schedule, and a time-phased project budget are all necessary inputs for forming a solid foundation for effective Milestone Cost Analysis. Additionally, any changes in the project scope that have not been fully comprehended in the baseline plan have to be included before an analysis is performed to create an accurate cost performance determination.

Set Up and Track Milestones

Using the Cost Baseline (Time-phased Budget), draw a planned cost performance curve that will be the baseline, annotating the milestones (Figure 10.8). This is a cumulative curve, typically expressed in resource hours or monetary units. An estimated number of hours are budgeted for each milestone, and because of the cumulative nature of the curve, when a milestone is reached, the cumulative number of hours for the milestone and all preceding milestones should be consumed.

As the project unfolds, actual cost data are collected and used to draw a cumulative actual cost curve, but what really matters is when a milestone is accomplished and marked on the actual curve. Hence, all performance is measured on the milestone level, following a fixed formula of 0/100 EVA measurement method. When the work on a milestone is started, 0% of the milestone's budget is earned, while the completion of the milestone earns a full 100%. Because of the cumulative nature of curves, the milestone acts as the culmination point of all previous project work, making its performance equate with project performance at that point.

Evaluate Project Results

This step involves comparing actual results of performing the project against its plan. The goal is to establish the schedule variance (SV) and cost variance (CV), and to identify the cause of the variances, if any. A good way to prepare is to do a rehearsal with milestone owners, and then to implement the evaluation in a progress meeting.

Periodic progress meetings instill the discipline and regularity in reviewing strides that are being made on the execution of a project. They should be called on a regular basis—once a month for a long project or once a week for a short project, for example. Each of the milestone owners is required to attend the meeting and report milestone cost management progress. The project manager should focus on asking the following five questions to gain a sense of cost management status:

1. What is the variance between cost performance baseline and the actual project cost?
2. What are the issues causing the variance?
3. What is the current trend and current prediction of cost completion if we continue with our current performance?
4. What new risks have been discovered and may affect the predicted cost at completion?
5. What actions should be taken to bring cost performance back in line with the baseline?

In our example from Figure 10.8 the variances are as follows:

$$\text{schedule variance} = \text{planned} - \text{actual}$$

$$\text{schedule variance} = 2 \text{ months} - 3 \text{ months}$$

$$\text{schedule variance} = -1 \text{ month}$$

and

$$\text{cost variance} = \text{planned} - \text{actual}$$

$$\text{cost variance} = 600 \text{ hours} - 900 \text{ hours}$$

$$\text{schedule variance} = -300 \text{ hours}$$

While the negative variance indicates that the project has fallen behind its plan, a positive variance means the project is ahead of its plan. No variance implies the performance is right on plan. Therefore, in our example, the project is one month late and 300 hours over the budget.

Predict Final Results

Certainly the most important and also the toughest step is the prediction of the final results. It is important because it enables a predictive look at the direction and trend of the project—where are our final cost and completion date going to end up? The absence of formulas for prediction such as those used in EVA makes the prediction an intuitive, challenging assignment, typically performed in the progress meeting. Such an exercise is very similar to one described in detail in the "Milestone Prediction Chart" section in Chapter 14. In particular, as the owner of the milestone describes its actual progress, potential variance from the baseline, and issues causing the variance, owners of dependent milestones evaluate the ripple effect of the milestone on subsequent milestones. The ripple effect is analyzed in the context of the critical path schedule, indicating the dependencies between milestones and related tasks. As a result of the analysis, predictions of milestones' cost and completion dates are made, all the way to the end of the project. If the final results are not favorable, corrective actions are developed to alter the trend and set the project back on track.

Using the Milestone Cost Analysis

Milestone Cost Analysis is a good candidate for both smaller and larger projects. With its visual power and little time to develop, the analysis serves well the needs of projects with smaller budgets. In larger projects, its primary rationale for use is its ability to supply a summary view of the project status to high-level managers, focusing on major project milestones.

With a bottom-up project plan already in place, a well-versed project team should take no longer than 30–45 minutes to perform a Milestone Cost Analysis that includes five or six milestones. As the number of milestones increases, so will the necessary time for the analysis.

CHOOSING YOUR COST MANAGEMENT TOOLS

The tools presented in this chapter are designed for different project situations. Project managers living under time pressures often ask, "Given my project situation, which are more appropriate to use?" To decide, take a look at the set of project situations given in Table 15.4, which lists various project situations and identifies which tools are geared for each situation. Consider this table as a starting point, and create your own custom project situation analysis and tools of choice to fit your particular project management style.

Situation	Budget Consumption Chart	Earned Value Analysis	Milestone Cost Analysis
Table 15.4: A Summary Comparison of Cost Control Tools			
Small and simple projects	✓		✓
Large and complex projects	✓	✓	✓
Formal progress reviews		✓	✓
Short time to train how to use the tool	✓		✓
Focuses on exceptions		✓	
Provides early-warning indication		✓	✓
Integrates scope, cost, and schedule		✓	✓
Provides a single control system for all levels		✓	
Takes little time to apply	✓		✓
Uses dollars or hours	✓	✓	✓
Displays trend		✓	✓
Provides built-in predictive approach		✓	
Little time available for cost management	✓		✓

References

1. Fleming, Q.W. and Koppelman, J.M. (2010). *Earned Value Project Management*, 4th ed. Newton Square, PA: Project Management Institute.
2. Christensen, D.S. and Heise, S.R. (2011). Cost performance index stability. *National Contract Management Association Journal* 25: 17–22.
3. Lock, D. (1990). *Project Planner*. Hunts, England: Gower Publishing.

16

PERFORMANCE REPORTING

Project performance reporting involves collecting key data concerning the performance of the project at a particular point in time, synthesizing the data into meaningful information, and then communicating the performance information to project stakeholders. Effective project performance reporting supports two key components of successful project management—open and strong lines of communication and transparent communication of information.

The intent of performance reporting is three-fold: (1) to determine and communicate how well the project is performing against the plan and baselines, (2) to use the information to enable informed project decisions, and (3) to develop forecasts and determine preventive and corrective actions as needed.

The tools presented in this chapter are intended to assist project managers in effectively fulfilling this three-fold intent. Just as performance reporting is a required activity of all project managers, the following performance reporting tools should be a part of every project manager's PM Toolbox. We begin by looking at the Project Reporting Checklist.

PROJECT REPORTING CHECKLIST

There are many types of project status reports with varying types of information contained in them. This is due in large part to variations in reporting needs and desires of project stakeholders, variation of industry reporting standards and practices, and variations in the size and complexity of projects. The project status report design and content is therefore situational, but should be standardized and consistent with the needs of the project stakeholders who are on the receiving end of the report. How does one establish and maintain standard messaging?

A simple and effective tool to consider is the Project Reporting Checklist. The checklist assists the project manager in determining the correct status information to include in a report and to consistently provide the information over time.

Project Management Toolbox: Tools and Techniques for the Practicing Project Manager, Third Edition. Cynthia Snyder Dionisio, and Russ J. Martinelli.
© 2025 John Wiley & Sons, Inc. Published 2025 by John Wiley & Sons, Inc.

Developing the Project Reporting Checklist

To be effective, a project status report must be current, concise, accurate, and contain only the information needed to keep stakeholders abreast of progress and the resources used to accomplish the project's objectives.

The Project Reporting Checklist will be different for every organization because every organization has its own unique set of information required for project reporting. Developing a standard set of checklist items is good practice, as it drives consistency in project reporting format and content within an organization.

The items contained within the checklist are developed by first understanding the information required by the project sponsor and other key project stakeholders. Then, additional items can be included that are unique to a particular project or to the project's second tier stakeholders.

Table 16.1 illustrates a sample set of project reporting items to consider as a reference and starting point for developing your own customized Project Reporting Checklist. The checklist shown is somewhat extensive, so keep in mind that the best project status report is concise and to the point. Developing your custom checklist will involve using a subset of items shown in Table 16.1.

The "status" column can be used to indicate that the information needed to develop a project status report has been collected. Some project managers add an additional column to the checklist labeled "source" to indicate where the source of information resides, or to provide a hyperlink to the source data itself.

Table 16.1: Example Project Reporting Checklist

Status	Checklist Items
Project Scope	
☑	Have the project objectives been changed?
☑	Have the deliverables in the project plan changed?
☑	Have there been any changes to the project scope?
☑	Are there any scope changes awaiting approval?
Project Schedule	
☑	Has the schedule been updated?
☑	Is the project progressing on the critical path, critical chain, or release plan?
☑	Does the time expended to date vary from the baseline schedule?
☑	Do you have adequate resources to maintain the schedule?
☑	Are project subcontractors or partners on schedule?
☑	What is the estimated completion date?
Project Budget	
☑	Has the budget been updated?
☑	Is the available budget to date in alignment with the budget baseline?
☑	What is the average monthly budget burn rate?

Status	Checklist Items
Project Performance	
☑	Have all deliverables been met to date?
☑	Have all project milestones been met to date?
☑	What is the earned value (EV)?
☑	What is the planned value (PV)?
☑	What is the actual cost (AC)?
☑	What is the schedule variance (SV)?
☑	What is the schedule performance index (SPI)?
☑	What is the cost variance (CV)?
☑	What is the cost performance index (CPI)?
Issues	
☑	Are there any current issues that need to be reported?
☑	What are the resolution plans for any open issues?
☑	Do your subcontractors or partners have any current issues?
☑	Do any issues require project sponsor or top management action?
Risks	
☑	What are the top 5–10 risks?
☑	What are the risk response plans for the top risks?
☑	What is the overall risk profile of the project?
☑	Do any risks require project sponsor or top management action?
General	
☑	Is the project being impacted by any external factors?
☑	Are there any quality issues associated with the project outcomes?
☑	Are you receiving payments as planned?
☑	Are there any actions or decisions needed on the part of the project sponsor or top management?

Using the Project Reporting Checklist

Most organizations have a defined point in time when project performance reporting is expected to begin. This can be as early as formal project initiation, at the project kickoff, or when the baselines are approved. Whenever the point, the project manager should begin using the checklist to formulate the content that will be included in the project reports.

In practice, the content included in a project report is fairly repetitive over time. However, it is valuable to review the checklist periodically to serve as a memory jogger to provide additional information in a report which may not be repetitive in nature. This is normally the type of information included in the "general" section of the checklist.

Additionally, different information may be required by the project stakeholders as a project progresses through the various stages of the project cycle. A review of the Project Reporting Checklist during these stage transitions will help the project manager modify the reporting content accordingly.

PROJECT STRIKE ZONE

Project performance reporting involves understanding and communicating how well project performance is progressing toward achieving the project objectives. Many times, project managers become over-focused on progress against their cost and schedule baselines, and forget that the real intent of a project is to achieve the business objectives driving the need for the project.

The Project Strike Zone is an excellent tool for evaluating and communicating progress toward achievement of the *project objectives*. It is used to identify the critical objectives for a project, to help a project manager and his or her stakeholders track progress toward achievement of the key business results anticipated, and to set the boundaries within which a project manager and team can operate without direct management involvement.

As shown in Figure 16.1, elements of the Project Strike Zone include the project objectives, target and threshold values, an "actual" field that provides indication of

Project Strike Zone				
Project Objectives	**Strike Zone**		**Actual**	**Status**

Value Proposition
• Increase market share in product segment
  • Order growth within 6 months of launch
  • Market share increase after 1 year
	Target
	
	10%
	5%
		Threshold
		
		5%
		0%
			
			7% (est)
			4% (est)
				Green

Time-Benefits Target • Project Initiation approval • Business case approval • Integrated plan approval • Validation release • Release to customers	1/03/2025 6/01/2025 8/06/2025 4/15/2026 7/15/2026	1/15/2025 6/30/2025 8/20/2025 4/30/2026 8/01/2026	1/04/2025 6/01/2025 8/17/2025 6/29/2026 TBD	Red
Resources • Team staffing commitments complete • Staffing gaps	6/30/2025 All project teams Staffed at minimum level.	7/15/2025 No critical path resource gaps	7/1/2025 Staffed	Green
Technology • Technology identification complete • Core technology development complete	4/30/2025 Priority 1 and 2 techs Delivered @ Alpha	5/15/2025 Priority 1 techs Delivered @ Alpha	4/28/2025 on track	Green
Financials • Program Budget • Product Cost • Profitability Index	100% of Plan $8500 2.0	105% of Plan $8900 1.8	101% est $9100 est 1.9 est	Yellow

Figure 16.1: Example Project Strike Zone

where a project is operating with respect to the target and threshold limits, and a high-level status indicator.[1]

A senior project manager for a leading telecom company described the culture within his company this way: "Managing a project is like having a rocket strapped to your back with roller skates on your feet, there's no mechanism for stopping when you're in trouble." Sound familiar? The Project Strike Zone is such a mechanism that is designed to stop a project, either temporarily or permanently, if the negotiated threshold limits are breached. At which point the project is evaluated for termination or replan and continuation.

Developing the Project Strike Zone

Developing an effective Project Strike Zone is a critical activity for ensuring that the project manager, project team, top management, and other stakeholders all understand and agree upon the objectives of the project. It is also critical for establishing the boundary conditions that will drive effective decision making on the project.

Defining a meaningful Project Strike Zone requires quality information from a number of sources. The initial set of objectives is derived directly from the approved Project Business Case (Chapter 2). To establish and negotiate the control limits for each objective, the project manager and the project sponsor need to know the project team's capabilities, experience and past track record, and balance thresholds against the new project's complexities and risks accordingly.

Identify Project Objectives

Identification of the project objectives begins during the initiation stage of a project. The factors represent a subset of the metrics normally tracked by a project team. The Project Strike Zone should include only the measures that represent the high-level project objectives (often the business objectives).

The strike zone is most effective when the objectives identified are kept to a critical few ones (usually 5–6), as this focuses the project and senior management's attention on the highest priority contributors to the success of the project. The factors deemed as "must haves" often include market, financial, and schedule targets, and value proposition of the project output.

Set the Target and Threshold Values

The target and threshold control limits shown in Figure 16.1 form the strike zone of success for each project objective. The target value for an objective should be able to be pulled directly from the project business case.

The threshold values represent the upper or lower limit of success for the project objectives.[2] Some discussion and debate is usually required to get an understanding of how far off target an objective can range and still constitute success for the project. For example, the target project budget may be set to $500,000. But, if additional spending of 5% is allowable, then the budget threshold can be set at $525,000. This means that even though a project team misses the target budget of $500,000, they are still successful from a project budget perspective if they spend up to $525,000.

Negotiate the Final Target and Threshold Values

Generally, the project manager establishes the recommended target and threshold values for each project objective, and then presents the information to the senior executive sponsoring the project. Based on the complexity and risk level of the project, and the capability and track record of the project manager and team, the project sponsor may adjust the values accordingly.

For example, on a project that is low complexity, low risk, and is being managed by an experienced project manager, the range between target and threshold values may be wider to allow for a higher degree of decision making for the project manager. Conversely, on a project that is of higher complexity, risk, or is being managed by an inexperienced project manager, the range between target and threshold values will be narrower to limit the decision-making authority of the project manager, at least initially.

Once the targets and boundaries are negotiated, the team should conduct the project, striving to stay within the strike zone threshold values.

Using the Project Strike Zone

Project managers utilize the Project Strike Zone to formalize the critical project objectives for the project, to negotiate and establish the team's decision boundaries with executive management, to communicate overall project progress and success, and to facilitate various trade-off decisions throughout the project life cycle (see "When Things Go Bad").

Executive managers utilize the Project Strike Zone to ensure a new project's definition supports the intended business objectives, and to establish control limits in order to ensure that the project team's capabilities are in balance with the complexity of the project. When used properly, it provides top managers a forward-looking view of project alignment to the business objectives. When problems are encountered, the tool's structure is intended to provide an early warning of trending problems, followed by a clear identification of "show-stopper" conditions based on the level of achievement of the project objectives. If a project is halted, senior executives can either reset the project objective targets or thresholds, modify the scope of the project to bring it within the current targets, or if necessary, cancel the project to prevent further investment of resources.

Executive managers and the project governance body approve the boundary conditions (targets and thresholds) of the Project Strike Zone between which the project manager can operate, thereby empowering the project manager to make decisions and manage the project without direct top management involvement. As long as the project progresses within the strike zone of each project objective, the project is considered on target and the project manager remains fully empowered to manage the project through its life cycle. However, if the project does not progress within the strike zone of each project objective, the project is not considered on target and the top managers intervene.

When Things Go Bad

Santiam was the code name for a multi-million dollar new product development project within a leading consumer electronics company. One of the primary strategic goals of the company was to move into a market outside of their traditional business. The product to be produced by the Santiam project team was the first introduction into the new market. To be successful, introducing the product into the market at the correct time, capturing a portion of the market share, and selling the product at a better price than their competitors were all critical project objectives needed to achieve the strategic business goal of entering the new market.

As most project managers know, however, the best developed plans are not immune to risks and alternative realities associated with doing business in a dynamic environment—the Santiam project was no exception. Six months into execution, word came from a supplier of a key component that was experiencing technological difficulties. One of the intermediate schedule dates identified in the Project Strike Zone was in jeopardy.

During the next project review with her executive leadership team, the Santiam project manager presented the updated Project Strike Zone with status on the product introduction date criteria presented as YELLOW (caution to management). Details of the current issue were discussed and risks associated with achieving the success criteria were reviewed. A mitigation recommendation to place a representative from the quality organization at the supplier's location to continuously monitor the situation and assist with solutions was approved by the executive team. The Santiam project team was given the go-ahead to continue development of the product, but under a heightened state of risk awareness.

After an additional three weeks it became clear that the problem had become a critical issue for the project when the supplier announced a six week slip in delivery schedule. This six week delay would cause a significant delay of the product into the market, turning the product introduction project objective in the Project Strike Zone from "YELLOW" to "RED" - meaning the project needed immediate top management intervention. An analysis of the other project objectives showed that a delayed launch would jeopardize the desired market share capture and drive the profitability index below the acceptable threshold value of 1.8 (see Figure 16.1).

In effect, the business case for the project was in the red zone. The project manager had the information she needed to make a recommendation to her project sponsor. Her recommendation was to cancel the Santiam project to prevent significant future losses to the business. Project discontinuance decisions are never easy, especially when thousands, or in this case, millions of dollars have already been spent. In the end, Bingham's executive leadership team utilized the information in the Project Strike Zone as the basis for their decision to cancel the project.

Use of the Project Strike Zone fosters a "no surprises to senior management" behavior by increasing the flow of relevant information between the project team and top management. This results in an efficient means of elevating critical issues and barriers to success for rapid decision making and resolution.

PROJECT DASHBOARD

In today's frenzied pace of many projects, project managers need to understand how the project they are responsible for is performing with respect to the key performance indicators, but they rarely have sufficient time to read through a number of detailed status reports from their functional teams.

Much like the dashboard of an automobile provides the driver a quick snapshot of the current performance of the vehicle, the Project Dashboard provides the project manager an up-to-date view of the current status of his or her project.

Unlike the Project Strike Zone which focuses on performance against the higher-level project objectives and business goals, the Project Dashboard focuses on the current state of the lower-level key performance indicators (KPIs). KPIs are the critical quantifiable indicators of progress toward an intended result. They focus on what matters most and are used for data-driven decision making.

The dashboard should be designed as an easy-to-read and concise (often a single page) representation of all KPIs as illustrated in Figure 16.2.

There are many types of Project Dashboards in use and available as reference for designing your own customized dashboard that represents the information most relevant and critical to your project. We like the design of the dashboard shown in Figure 16.2 because its graphical nature provides a variety of project status measures, and it is concise.

Designing a Project Dashboard

The Project Dashboard is one of the most flexible and customizable tools in a project manager's toolbox. As stated earlier, it needs to be designed around the particular key performance indicators of a project. Since each project is unique, each project will have somewhat unique performance indicators and therefore will likely have a unique Project Dashboard design.

Identify the Key Performance Indicators

Design of the Project Dashboard begins with identifying the key performance indicators for the project. These typically can be found in the Project Business Case or Project Charter (Chapters 2 and 3).

The project objectives, identified and quantified in the Project Strike Zone, define the end state of the project in terms of what value the project brings to the sponsoring organization. The KPIs quantifiably measure how well the project is performing toward accomplishment of the project objectives.

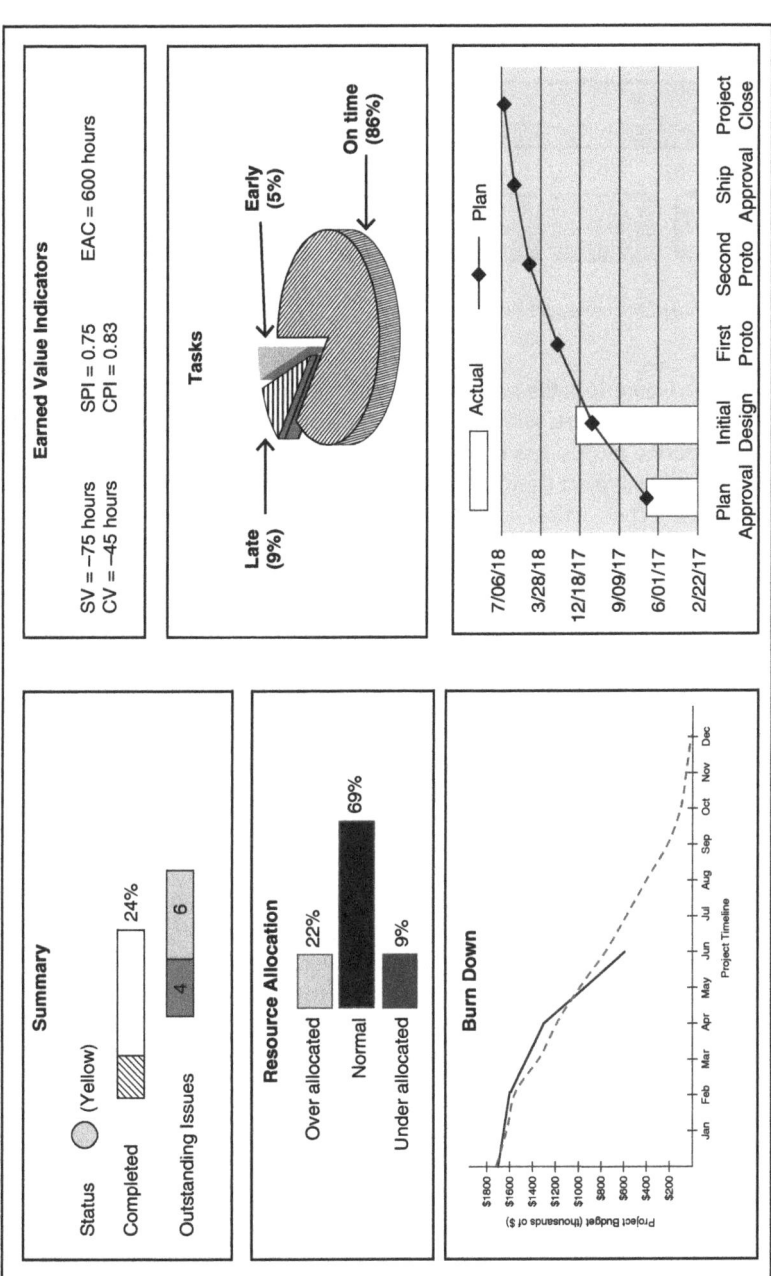

Figure 16.2: Sample Project Dashboard

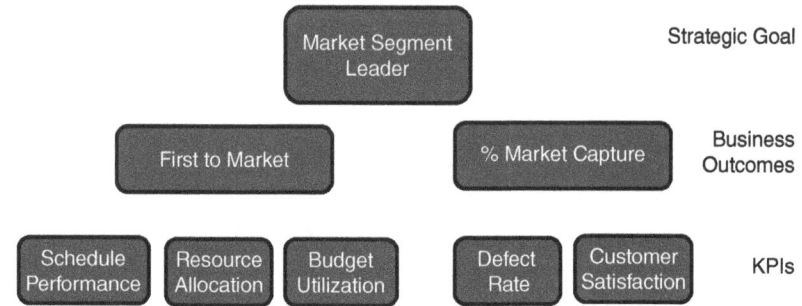

Figure 16.3: Strategy, Outcomes, and KPIs

As shown in Figure 16.3, the project KPIs are part of a measurement hierarchy that must be understood. It starts with the strategic goals the business wants to achieve. These are supported by business outcomes, which are in turn supported by KPIs. The KPIs identified in the Project Dashboard should directly measure performance toward achieving the project objectives documented in the Project Strike Zone.

For example, a strategic goal for an enterprise could entail being the leader in a particular market segment. Business outcomes that support the strategic goal would be first-to-market with new products and an established market share percentage. Project objectives would need to quantifiably define the project completion date that ensures first-to-market position for the project outcome. Three project KPIs that support first-to-market include (1) schedule performance, (2) resource allocation percentage (if resources are not close to 100% allocated to plan, schedule will likely suffer), and (3) budget utilization (ensuring that the project doesn't run out of funds). KPIs that support percent market capture, defect rate, and customer satisfaction.

Outline the Dashboard Layout

Based on the key performance indicators selected in the previous step, you have an idea of what information should be shown on your dashboard. Now you need to determine *how* you want to present that information on the Project Dashboard.

To accomplish this, take a few minutes to sketch the structure of the dashboard as shown in Figure 16.4. Nothing fancy here, just sketch out the location of the information on the page. The goal is to design the dashboard so it is both comprehensive in content and is appealing to the eye of the recipient.

Populate the Dashboard

The final step in designing the Project Dashboard involves locating the pertinent performance data and representing it on the dashboard. Whenever possible, use graphical representations as they facilitate a quicker analysis of the current performance on the part of the recipient than a text-based representation.

Some project managers embed hyperlinks within the top-level performance graphics that link to detailed data about the KPI of interest. For instance, if additional detail is

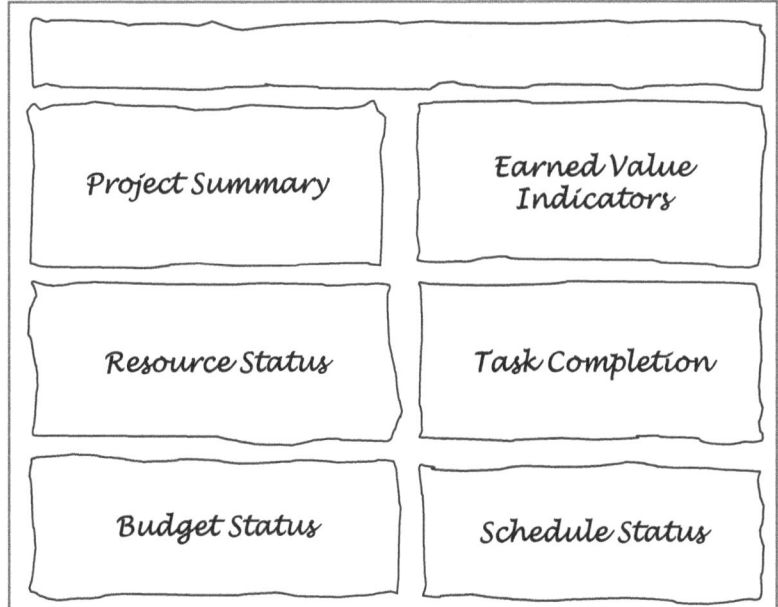

Figure 16.4: Sample Dashboard Structure Layout

needed for the *schedule performance* KPI, a link can be provided to a detailed Gantt Chart, Milestone Analysis Chart, or even the schedule section of the current project status report.

Using the Project Dashboard

The Project Dashboard can be used as both a communication tool and decision support tool by project managers. By using the Project Dashboard to synthesize lower-level performance data into higher-level information, a project manager becomes armed with the right information he or she needs to communicate the current status of the project with respect to the KPIs. Additionally, many decisions have to be made during the course of a project, some large and some small, and the Project Dashboard serves as the basis of information to make decisions (see "Tips for Using Project Dashboards").

The Project Dashboard is used to consolidate and display performance information that resides in various project data sources. For example, schedule performance data may reside in a Gantt Chart, spend rate data in a Burn Down Chart, and cost performance data in the Earned Value Management system. The dashboard becomes a single source of key performance information for a project.

The Project Dashboard also serves as a data source for the development of an overall project status report. The project manager can use the dashboard as the data source for the performance against the KPIs information that is normally included as part of the Summary Project Report.

Tips for Using Project Dashboards
With the overall simplicity of a Project Dashboard, project managers need to remember that dashboards are not, in and of themselves, a panacea. The dashboard is only as effective as the design of its structure, the value of the measures and metrics chosen, the accuracy of the data represented, and how effectively the dashboard is used to drive communication and decisions.

With the overall simplicity of a Project Dashboard, project managers need to remember that dashboards are not, in and of themselves, a panacea. The dashboard is only as effective as the design of its structure, the value of the measures and metrics chosen, the accuracy of the data represented, and how effectively the dashboard is used to drive communication and decisions.

Project managers must avoid descending into a quantitative and analytical quagmire when using dashboards. There is a real return on investment that must be maintained in that the value gained from the use of the Project Dashboard must be greater than the cost of obtaining and analyzing the information contained within it.

Beware of false and conflicting information that may show up in a dashboard. Take the time to ensure that the information is current, accurate, and that it conveys the right message about the project performance against the KPIs. If not, the dashboard may do more harm than good.

The Project Dashboard helps the project manager focus on the key performance indicators and how the project is operating relative to the indicators. It supports the project manager in capturing and reporting specific data points relative to the key performance indicators.

SUMMARY STATUS REPORT

The Summary Status Report is a document that highlights and briefly describes the status of the project, reporting on the scope, cost, and time variance, showing significant accomplishments, identifying issues, predicting trends, and stating actions required to overcome issues, risks, and reverse negative trends. A sample of a Summary Status Report is illustrated in Figure 16.5.

Developing the Summary Status Report ensures a proactive, predictable cycle of project control, communicating information about project problems and status to all concerned stakeholders, including top management, and taking actions to put the project on track.

Developing a Summary Status Report

Producing a meaningful Summary Status Report starts with quality information inputs. A solid set of project baselines, such as scope, schedule, and cost, is the foundation for a good progress report. The baselines are compared to the actual state of the project to assess its performance. The actual state is derived from work results and other project records. Through work results, for example, we report which tasks or deliverables are completed and resources expended, presenting them by means of schedule and cost

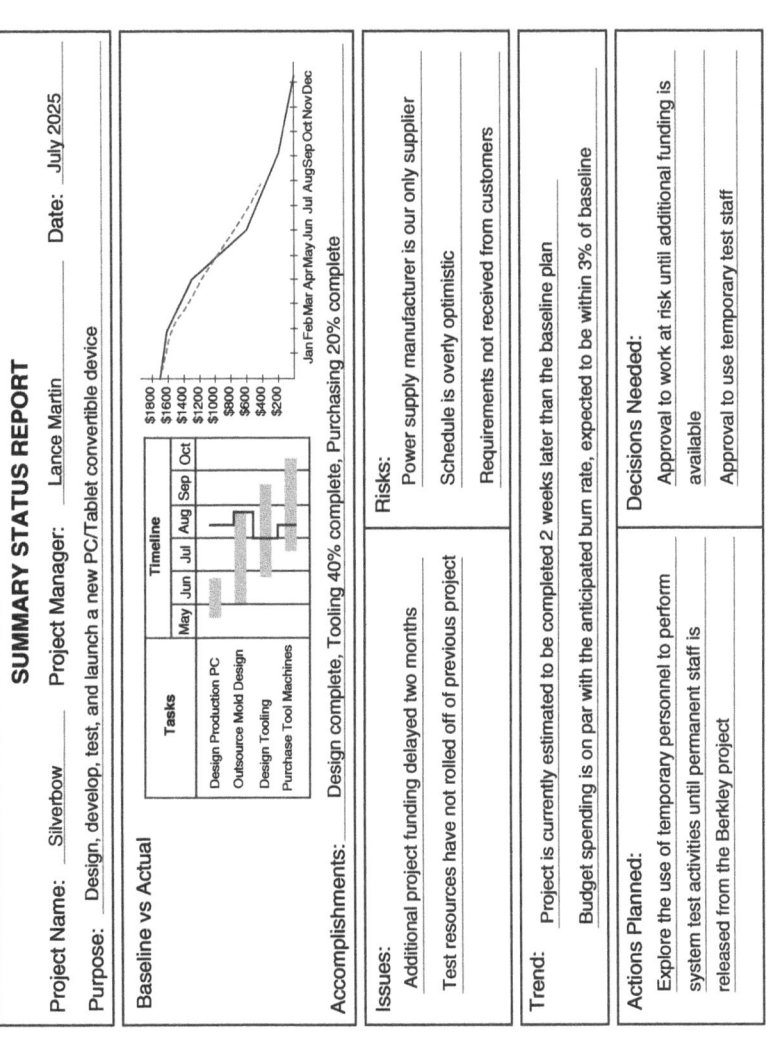

SUMMARY STATUS REPORT

Project Name: Silverbow   Project Manager: Lance Martin   Date: July 2025

Purpose: Design, develop, test, and launch a new PC/Tablet convertible device

Baseline vs Actual

Accomplishments: Design complete, Tooling 40% complete, Purchasing 20% complete

Issues:

Additional project funding delayed two months

Test resources have not rolled off of previous project

Risks:

Power supply manufacturer is our only supplier

Schedule is overly optimistic

Requirements not received from customers

Trend:

Project is currently estimated to be completed 2 weeks later than the baseline plan

Budget spending is on par with the anticipated burn rate, expected to be within 3% of baseline

Decisions Needed:

Approval to work at risk until additional funding is available

Approval to use temporary test staff

Actions Planned:

Explore the use of temporary personnel to perform system test activities until permanent staff is released from the Berkley project

Figure 16.5: Sample Summary Status Report

control tools such as the Jogging Line and Earned Value Analysis. Other information describing the project execution may be included in project records such as change requests and issues.

Due to the unique nature of projects and the organizations that perform them it is useful to define the information that will be reported, the format, frequency, responsibilities for collecting the information, and the distribution list. The design of the report needs to provide just enough information to provide both a summary overview of past progress and projections for the future. Reports that are too detailed and too onerous to fill out are not a good use of project manager's time. See the section on The Case of Over Reporting.

The Case of Over Reporting

This is a progress reporting story of a project manager in an enterprise IT department that, unfortunately, is all too common. According to the project manager, "project managers in our group write a project status report every month for every project they manage. I am expected to show how much time I spend on each project, including all administrative work. Frankly, every project manager just reports 100% of their regular work hours, even though we may be working 120% or 130% of our regular work hours, just to minimize the amount of work we have to report on.

These are really very long reports, almost always 6–7 pages. I usually manage four to five projects at a time and do a report for each of them. It takes a lot of time to write them and frankly, it is time I should be using to manage my projects. I have a hard time believing that our managers really spend time reading 20–25 detailed reports each month."

This is a case of over reporting in which value is lost for both the project managers and their leaders. A better approach would be to provide a Summary Progress Report that would add more value to everyone involved.

Determine the Variances

The key to a meaningful summary status report is turning the project data being collected into useful information about the performance of the project. Begin by determining the variance between the baseline and the actual project status.

How the variance is collected is situational and involves a number of other tools. If for example the project is a small departmental project, it might use the Jogging Line to indicate the schedule variance. Such a non-project-driven department might easily opt to show only this type of variance. In a different situation, where this would be a large project to design and deploy performance metrics throughout a project-driven firm, the Milestone Analysis might be used to identify the schedule and cost variance, or even the full-scale Earned Value System might be employed and supported by verbal descriptions about the quality and scope variances.

Ensure that you are reporting progress at the project level. A good strategy is to use the WBS as a framework, as described in the WBS section in Chapter 6. The process of reporting starts at the work package level, identifying the variances, and aggregating them up the WBS hierarchy to establish the variances for the whole project. Subsequent steps of identifying issues, predicting trends, and specifying corrective actions should follow the same approach of using the WBS as the framework.

Identify Issues and Risks

If there is a variance, especially an unfavorable one, report the issues causing the variance. Working with issues looks at the present problems that are at the root of the variance and what their impact on the project is.

Generally, issues impacting the project progress may come from any area of work. It may be useful to use an Issue Register to track all project issues. Issue Registers are an effective way of tracking problems on the project. The issues that are impacting the project, or those that need top management involvement to resolve, are carried forward to the Summary Status Report. Figure 16.6 shows a sample Issue Register.

As discussed in Chapter 10, you will want to identify the risks that may occur in the future and report the impact they will have on the project if they do occur. For instance, consider a project where the project manager just learned that one of their major materials suppliers might be on the verge of bankruptcy. This possible event would be identified as a high-level risk, worthy of reporting on the Summary Status Report. The team would immediately develop a response strategy instead of waiting to hear a few months later about the bankruptcy, at which point they might be helpless to correct the impact on the project.

Predict Trends

This section of the report shows the predicted future performance based upon the current status of the project. Although forecasts of this type are not easy and are notoriously vulnerable, their essence is less in their accuracy and more in their creation of early-warning signals. For example, in the Summary Status Report in Figure 16.5 "project is currently estimated to be completed 2 weeks later than the baseline plan" is a clear warning, one that mandates action to attempt to reverse the trend. The ability to forecast the trend, week after week, or whatever the report frequency, is paramount in building an anticipatory climate where project teams are alert about the project's past progress but even more about what the future bears.

Project Issue Register					
Issue #	Issue Description	Date Raised	Owner	Priority	Status
1	Second round of funding not approved	10/8	Williams	1	Closed
2	Research data not available until January	10/25	Owens	1	Open
3	Quick set-up feature broken	10/26	Powers	2	Open
4	Vendor missed 1st delivery	11/23	Gupta	1	Closed
5	.....				
6	.....				

Figure 16.6: Sample Issue Register

Specify Actions

If trends are unfavorable, this section identifies corrective or preventive actions that should be taken so you can deliver to the baseline plan. With a look at the trend section, we need to specify corrective actions, assess their impact, and assign an owner in the report. Along with the trend, the specification of corrective actions is perhaps the most valuable part of the report because it enables project teams to be proactive. While the performance progress is important in telling where we are, it is no more than the project history—there is really nothing that we can do to correct or change it. Our only opportunity to change the project is in the future, and that is what the trend and corrective actions offer: an opportunity to anticipate and shape the future by acting now.

Using the Summary Status Report

Whether small or large, projects need the Summary Status Report. Pressed for resources, small projects—especially in a multi-project environment—will likely issue the summary report as their only report, doing away with a detailed report.

Although many will prepare the report in a formal, written format per predetermined frequency, it is not unusual for managers of small projects to report status verbally (see "The Case of Underreporting").

The Case of Underreporting

We heard this story during a ten-minute lunch with a project manager for a technology firm. According to the PM, "we develop components for our internal customers who build them into their new products for external customers. With seven projects that I am managing right now, I don't really have time to write progress reports. This is really the case for all the project managers. All of us run multiple projects at a time, too many we believe, and no one has time for reporting since we typically work 70 hours a week.

My boss would like to have the reports, but knowing how busy we are, he doesn't require them. He was in our shoes before he was promoted to this position, so I guess he understands what kind of situation we are in. He does ask us in our weekly staff meetings if we have any problems he can help with. But he can't really help much because he has no resources to help out. I usually develop a Gantt Chart for each project, but with this pace of work, I just don't have time to keep them updated."

This is a case of severe under reporting, and a dangerous situation to be in. This project manager, his manager, and their organization will be in a continuously reactive or "fire-fighting" situation without the use of a streamlined Summary Status Report for each project.

An hour may be sufficient to prepare a typical Summary Status Report for a small- or medium-sized project. Even as time requirements go higher with the size and complexity of projects, it is clear that a few hours of a large project team's time should suffice for the summary report production. This assumes that extra time—perhaps running in tens of hours—was spent to generate the performance data that feeds into the report.

PROJECT INDICATOR

There is a direct correlation between the Summary Status Report in the previous section and the Project Indicator. The Project Indicator is a presentation device to communicate the information contained in the summary report. The Project Indicator gives the project manager a high-level view of the total project and helps him or her to determine if the project remains successfully on track or if there are potential barriers and issues that must be addressed.

The Project Indicator functions as a key communication vehicle between the project manager and the top managers that highlights key cross-project issues that need to be elevated to senior leadership for resolution. It effectively facilitates focused discussion between the project manager and their managers. For project managers that are uncomfortable having these discussions with their senior leaders, or those that are new to the opportunity, the Project Indicator also provides a mental prop-and-cue card for discussion topics to cover.

It is effective to have a common Project Indicator format in use on all projects for consistency and comparability of information. The reporting format should include all critical project elements that are important to senior management so that they can quickly evaluate progress on projects and determine which need more of their focus and attention. A sample Project Indicator is displayed in Figure 16.7.

The Project Indicator is brief and limited to one or two pages. It is meant to give a concise, but comprehensive description of current project status, key issues and changes that have been encountered, performance against project performance metrics, and the management of critical risk events.

Developing the Project Indicator

Creating a Project Indicator begins with understanding the information that should be included and communicated in the informal or formal project review with top management. As stated previously, it is useful to top managers if all Project Indicators are consistent in form and content. This is probably another application of the 80/20 rule; 80% of the information contained in the Project Indicator should be common to all projects being evaluated by the organization's top management, and 20% of the information should be unique to each project. In general, the Project Indicator should include the following information:

- Significant changes to the project
- Work completed since the last review
- Work planned during the next reporting period
- Performance against plan
- Issues encountered
- Risks identified.

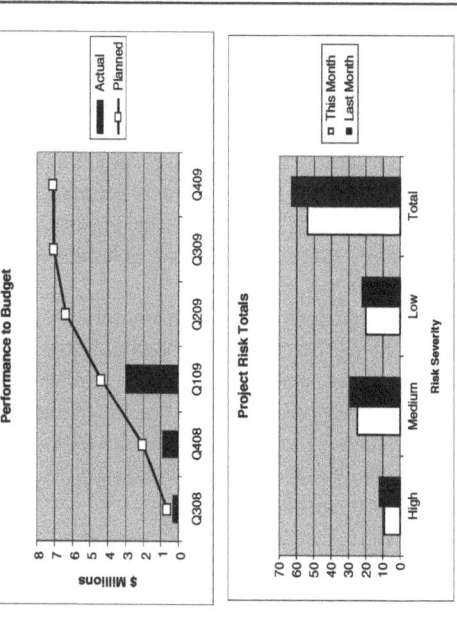

Project Performance Indicator

Project Changes

- Additional test cycle approved
- Second materials vendor approved

Project Overview

- Circuit board power on complete
- Currently 4 weeks behind schedule
- Currently $1.2 million under budget

Project Status

Work accomplished last month:

- Strategic customers identified and committed
- Circuit board power on complete
- SW build 42 delivered

Work planned for coming month:

- Validation team staffed and fully tasked
- Marketing plans completed
- Enclosure CAD files delivered to vendor

Current Issues

- Validation platform stability
- Four weeks behind schedule on evaluation
- Critical part shortage for next circuit board builds
- Currently a five week gap in engineering resources

Figure 16.7: Sample Project Indicator

Changes to the Project

A brief description of the significant changes to the project that may have an impact on performance should always be included in a Project Indicator. Example changes include significant scope increases or decreases, changes in project budget or funding, changes in project resources, and changes to the project objectives.

Project Overview

This section provides a few bullet points that summarize the overall status of the project. It can identify key deliverables that have been completed, the project status, and the budget status. Other sections of the Project Performance Indicator will provide the backup information for the Overview.

Project Status

This section of the Project Indicator describes work accomplished since the last report and work planned for the coming reporting period. The work accomplished section provides an overview of the key elements of work completed, such as the achievement of project milestones, completion of project deliverables, the resolution of major blocking issues, removal of risk events, and the completion of key project events such as a customer review or signing of a partnership agreement.

A high-level description of the work that is planned between the current reporting cycle and the next allows for a discussion on not just what has occurred in the past, but also what will occur in the near future, and what you as the project manager are anticipating in the next work cycle.

Performance Information

The performance information, such as performance to schedule and performance to budget, should provide a concise description of how the project is performing against the key performance indicators. This can be accomplished either graphically (which is always preferred when communicating to senior managers), or in text. If the project is using Earned Value Analysis (Chapter 15), those values should be included in this section.

Current Issues

This section should concisely describe the major issues that the project team is working to resolve. Remember that the Project Indicator is meant as a verbal communication tool, so issues are briefly described on the indicator and then explained to whatever detail is needed during the ensuing conversation with top managers. For every issue communicated on the Project Indicator, there must be an action plan that is being put in to play.

Risks

Much like the project issues, a concise description of the critical risk events should be included in the Project Indicator. Try to limit the risks to the critical three to five events that the project team is working to eliminate or mitigate. Even better, a summarization of progress against risks may be a better representation of project risks.

Other Items

Since each project is unique, include a section in the indicator that addresses information that pertains strictly to the project at hand. For reference on what to include in this section, the "General" section in the Project Reporting Checklist is a good source.

Using the Project Indicator

The Project Indicator is a verbal communication tool. It is used to communicate overall project status to the top management of an organization. In the process it also facilitates the necessary discussion between the project manager and his or her senior leaders. For each item included in the indicator, ensure you have the backing details to engage in a conversation, or include the right member of your project team who can speak to the details.

The indicator can become the means to engage top management in the critical aspects of the project and facilitate a request for assistance if and when needed.

The Project Indicator can also be used for effective communication of project status to the project team. Often, project team members are not privy to the overall status of the project they are a part of. To be most effective when used in this manner, the project team should be briefed *after* top management so pertinent aspects of the conversations with management can also be communicated to the team members.

To provide a complete overview of project status, many project managers present the Project Indicator along with the Project Strike Zone. This provides a more holistic message that incudes operational status with review of the project objectives and current performance against those objectives.

Variations

Some project managers use indicators within their project teams to facilitate intra-team progress reporting. In this use case, the functional project team leaders prepare and present a more detailed and focused Functional Indicator that reflects the work of each functional team (see Figure 16.8 for an example).

The project manager must work with each functional team leader to determine the best format and content to present in the Functional Indicator. By encouraging the functional leaders to keep their Functional Indicators current, the project manager will receive a comprehensive yet concise report on functional team status on a regular basis (recommended weekly). Additionally, the information in the Functional Indicators can be used as a data source for building the Project Indicator.

Software Development Project Indicator

Software Development Status

Work accomplished last week:
- Updated latest software build with User Guide

Work planned for this week:
- Debug critical bugs
- Complete Linux support plan

Next deliverables:
- Linux support plan—due 10/14
- Software build 42—due 11/01

Critical Bugs

1. Operating system is crashing
2. New firmware release causes system interrupts

Risks

Risk: Linux developer not rolling off Icon program when expected

Impact:
Schedule will be delayed two weeks.

Mitigation:
Borrow developer from the technical marketing team for short-term relief

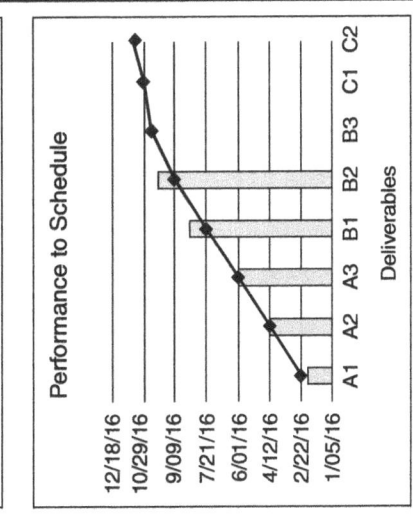

Software Bug Trend

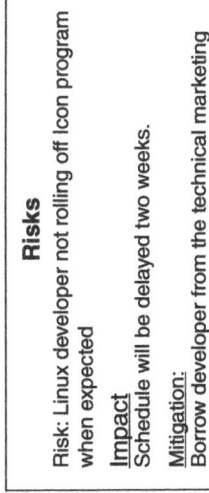

Performance to Schedule

Figure 16.8: Sample Functional Indicator

Table 16.2: Project Reporting Tools

Situation	Reporting Checklist	Project Strike Zone	Project Dashboard	Project Indicator	Summary Project Report
Prepare information to report	✓		✓		
Tailor reporting information based on project life cycle	✓				
Communicate performance against project objectives		✓		✓	✓
Facilitates project-level decision making		✓	✓	✓	
Communicate performance against operational KPIs			✓		✓
Communicate functional status to project manager			✓		
Communicate overall project status to top managers		✓		✓	✓
Describe current issues			✓	✓	✓
Communicate project trends and risks		✓	✓	✓	✓
Best for written status reporting					✓
Best for verbal status reporting		✓	✓	✓	

CHOOSING YOUR REPORTING TOOLS

The tools presented in this chapter are designed for various project performance reporting situations. Matching the tools to their most appropriate usage is sometimes a bit confusing. To help in this effort, Table 16.2 lists various performance reporting situations and identifies which tools are geared for each situation. Consider this table as a starting point, and create your own custom project situation analysis and tools of choice to fit your particular project management style.

References

1. Martinelli, R., Waddell, J., and Rahschulte, T. (2014). *Program Management for Improved Business Results*, 2nde. Hoboken, NJ: John Wiley and Sons.
2. Martinelli, Russ and Jim Waddell: *The Program Strike Zone: Beyond the Bounding Box* (Project Management World Today, March–April 2010).

17

PROJECT CLOSURE

This chapter focuses on best practice project closure tools, techniques, and functions the project manager facilitates to ensure project closure leads to continuous improvement from lessons and experiences learned. Four specific tools will be detailed in this chapter: Project Closure Plan and Checklist, Project Closure Report, Lessons Learned, and Postmortem Review. Prior to the tools, however, an overview of project closure is offered to ensure understanding with regard to closing a project.

Project closure consists of the work necessary to end, or close out, any pertinent project activities across life cycle phases and the project itself. Such work here is often associated with contractual obligations, procurement paperwork, handoff agreement(s) between the project and operations teams, and reallocation of resources, just to mention a few.

Project closure is not reserved for the end of the project, but rather closing is about reviewing and learning throughout the project life cycle. Waiting until the end of the project to reflect and learn becomes a problem not only for the project, but also for their entire company. Any project team that waits until the end of the project's life cycle loses out on opportunities to learn and improve throughout the project. Highly experienced project managers and organizations with mature project management organizations know the value of intentionally engaging stakeholders in closing work throughout the life cycle of any project.

The importance of closure is twofold. First is to ensure that the work planned was actually completed. If there is any variance between what was planned and what was completed, the project manager must determine why and then determine if any adjustments need to be made for the remainder of the project based on why the variance occurred. Variances here could be either positive or negative to the successful completion of the project.

The second aspect of closure is to determine what worked and what didn't, and then document the findings. Documentation here means updating project artifacts, planning documents related to upcoming phases of work, and templates and best practice documents for other project teams. Then the lessons learned must be socialized. Doing so shares knowledge, so that all project teams can plan and execute their projects more efficiently and effectively.

Project Management Toolbox: Tools and Techniques for the Practicing Project Manager, Third Edition. Cynthia Snyder Dionisio, and Russ J. Martinelli.

PROJECT CLOSURE PLAN AND CHECKLIST

Proper project closure starts with proper planning. Planning is especially important when it comes to project closure because this phase in the project's life cycle and the work associated with it is often neglected. To ensure project closure is not neglected, best practices suggest starting the project's closure work during the planning phase of the project. Doing so ensures allocation of resources and clearly sets expectations of how project closure activities will be conducted throughout the project.

There are three major tasks for project managers to oversee during close out work. First, the project manager must evaluate if the outcomes, decisions, and deliverables of the project (at that point in the life cycle) meet the expectations of all stakeholders. This work can be summarized by the following questions:

- Are all planned and scheduled deliverables and key milestones complete?
- Based on the work to date, is the overall health of the project and the team high and functional?
- Based on the work to date, are the stakeholders satisfied with progress and optimistic about the effectiveness of work relative to the next phase (or operationalization) of the project?
- Is all contract and procurement work finalized and up-to-date?
- Is the team adequately prepared, resourced, and optimistic about the effectiveness of work relative to the next phase of the project?

Essentially, these questions serve as a framework for project managers to begin project closure activities. Ideally, all questions would be answered, "yes." If there is a "no" answer, the project manager needs to further investigate, determine what is needed to change the "no" to a "yes" in the next phase of work. In some cases, doing so will require support, guidance, and direction from senior leadership, the governing body, or sponsor of the project.

As project closure activities begin, a multitude of work lies ahead. The use of the Project Closure Checklist is helpful for guiding the closure work and ensuring all areas are addressed. Table 17.1 offers a generic Project Closure Checklist that can be modified to meet the needs of your particular organization and project.

Developing the Plan and Checklist

To plan for project closure, you should include the core project management plan components, baselines, and other applicable logs, registers, and documents. Since every project is different, the detail (breadth and depth) of the project closure plan will be different for each project. The project manager must discern at which critical points in the project should an investment of time, personnel, and perhaps other resources be made to reflect, learn, update documents and plans, and share findings.

Table 17.1: Generic Project Closure Checklist	
Status	Checklist Items
☑	Verify all project deliverables completed
☑	Document actual delivery dates of all deliverables
☑	Document final change management log and final project scope
☑	Officially notify the customer or client of project completion
☑	Officially notify vendors, suppliers, and partners of project completion
☑	Close out all work orders, contracts, and subcontracts
☑	Submit final customer or client acceptance documentation
☑	Prepare and submit final reports (project, financial, quality, risk, and so on)
☑	Conduct the final post-project review meeting
☑	Collect, organize, and document all project lessons learned
☑	Recognize the work of project team members and celebrate success
☑	Ensure all costs are charged to the project
☑	Submit final invoices to customer and pay final invoices from suppliers
☑	Close out all financial documents
☑	Compile and store all required documentation for long-term data management
☑	Dispose of all equipment and materials

A project closure plan should include the following sections or categories of content.

- Executive Summary
- Project Scope and Business Objectives
- Start and End Dates of Project Phases of Work
- Project Completion Criteria and Metrics
- Project Closure Deliverables
- Project Closure Documentation
- Project Closure Resources
- Project Closure Communication Plan
- Final Approval.

It is important to note here that the closure plan will (at the completion of the project) turn into the basis of the project closure report. So, in essence, using the outline above enables the project manager to plan with the end in mind. The following describes each of the sections.

Executive Summary

Details the project closure work effort, outlines key findings, issues, best practices, recommendations, and lessons learned.

Project Scope and Business Objectives

Summarizes the scope of the project and the key business objectives. This section explains stakeholders and groups (inside and outside the organization) impacted by the project and aims to align deliverables and project work outcomes to business goals. It is important that any objective detailed in this section be plainly written, unambiguous, and specific to the point of being measurable. Tools such as the Project Business Case (Chapter 2), Project Charter (Chapter 3), and WBS or PWBS (Chapter 6).

Start and End Dates of Project Phases of Work

Summarizes the estimated start and end dates of the project. Some project managers may also choose to illustrate a milestone view of the project schedule in this section. Doing so helps to highlight the critical timing of the project work and at which points closure activities will be used. Being clear in this section about project closure events helps to ensure project resources (especially personnel) are available for such work prior to be reallocated to other initiatives.

Project Completion Criteria and Metrics

This section details both project and business measures and metrics that will be used to evaluate project completion. A simple table, as shown in Table 17.2, is most often used to denote the activity or deliverable being reviewed, the criteria used to determine completion, and any metrics used in the completion determination. Having this level of detail creates unambiguous means by which a project can be deemed successful.

Project Closure Deliverables

This section outlines two major closure needs. First, it serves as a checklist for review of all project deliverables and whether or not each has been completed. Second, it identifies the hand-off necessary to release project deliverables to operational owners. A simple table, such as the one shown in Table 17.3, is a simple way to capture such detail.

Table 17.2: Project Completion Template		
Activity/Deliverable	**Completion Criteria**	**Completion Metrics**

Table 17.3: Project Closure Deliverables Template			
Deliverable	Operational Owner	Complete	Handoff Plan

Any outstanding items such as those not complete or not yet ready for handoff will be accompanied with recommendations and actions for resolution.

Project Closure Documentation

Any project document should be reviewed for completeness, updated, and archived. To do so, a document or artifact register should be detailed at the planning phase of the project and updated throughout the life cycle of the project. This section of the plan should include that register, clearly noting the owner(s) during the project and post project, the location of the artifact, and keep track of the updates or versions of the artifacts as the project matriculates its life cycle.

Project Closure Resources

This section outlines the resources needed for closure activities. Planning for the use of resources early in the project helps to ensure their availability at the time of closing activities. Again, a simple table (see Table 17.4) will suffice for planning purposes.

Project Closure Communication Plan

This section details the communication necessary to broadcast all lessons learned, key takeaways from the project work, new best practices, and any updated tools, templates, or other artifacts. A simple table (see Table 17.5) outlining key aspects of the communication plan can help serve as a summary view of all messaging.

Table 17.4: Project Closure Resource Template		
Resource Name	Closure Activity Responsibility	Duration of Work Effort

Table 17.5: Project Closure Communication Template				
Owner	Message	Audience	Distribution Method	Timing

Final Approval

This section is a placeholder for final sign-off of project closure. It is a formality that is often signed singularly by the project sponsor or jointly by the sponsor and operational leader assuming responsibility of the project deliverables. A name, signature, and date are the items most often captured and once obtained signal the formal closure of the project and release of all resources.

Using the Closure Plan and Checklist

It is incumbent upon the project manager to use the Closure Plan and Checklists as tools to help negotiate the need for resource allocation for such work and to support the monitoring function of the project and expedite the review of work completion. The intent is to validate completion of work and readiness for handoff while simultaneously codifying lessons learned and sharing them for enterprise value. Use just enough rigidity necessary to capture this work, while maximizing latitude to project personnel necessary to complete the work.

PROJECT CLOSURE REPORT

Most project sponsors expect a final project report. When done properly, the Project Closure Plan (discussed above) manifests into the Project Closure Report. This tool is important in that it is the final report that formalizes the closure of the project. It confirms not only the completion of the project, but also the acceptance of the project handoff from the project team to the operational owner or client.

Developing the Project Closure Report

The Project Closure Report is a documented review of the entire project. It highlights the completion of project work along with any variance between what was planned and what was actually accomplished. Such variances may be in the form of schedule, cost, resource utilization, and other pertinent measures and metrics. It should also highlight the likelihood of application of the project outcome(s) achieving the initially noted business case problem or opportunity. Noting all of this can follow the same outline and format as the project closure plan, which is outlined again here for convenience.

- Executive Summary
- Project Scope and Business Objectives
- Start and End Dates of Project Phases of Work
- Project Completion Criteria and Metrics
- Project Closure Deliverables
- Project Closure Documentation
- Project Closure Resources
- Project Closure Communication Plan
- Final Approval.

In addition to the tables and checklists outlined in the prior section, you may choose to summarize major closure activities into an executive-level checklist. This executive-level view of the project would summarize the milestones, deliverables, key decisions, hand-offs, signoffs, and other pertinent aspects of the project.

Table 17.6 illustrates a sample executive-level checklist used in a Project Closure Report. The content and level of detail will likely vary from project to project, but this illustration provides a general view of the checklist.

In developing this report, the sponsor or stakeholders may request a presentation to accompany the final report. Therefore, as the report is being planned and detailed, be mindful about how the report may be distilled into a presentation format for use.

Using the Project Closure Report

A Project Closure Report is used to help project consider the information that the project sponsor and other stakeholders need at the closing of the project. The primary use is to provide a summarized document of the entire project for sponsor to sign-off for project completion. However, other stakeholders benefit as well. For the end-user or operational team, it is an official hand-off of ownership. For the project manager and team, it outlines a clear point of transition to work on other projects. For other project teams, there is value in gaining lessons learned and best practices as well as identifying any implications from the closure of the project.

Although the report can be planned at the beginning of the project, it is not until the end that it is fully documented to completion. In between planning and closing, the project manager must use discretion to periodically update the report. Doing so may expedite some time at the end. However, even if a decision is made to wait until the end,

Table 17.6: Project Completion Template		
Item	Complete (yes/no)	Notes or Actions
All deliverables complete		
All milestones complete		
All contracts closed out		
All issues resolved		
All hand-offs signed off		
All payments made		
All invoices submitted		
All accounts closed		
All artifacts updated		
All lessons learned gathered, documented, and shared		
All personnel reassigned		
All excess materials disposed of or stored		
. . .		

the project closing plan and checklists (outlined above) should be used throughout the project. At the end of the project, these checklists can be used to complete all necessary components of the closing report.

Closing a project is more than just handing off ownership to an operational team or customer. As a project manager, it is also about recognizing the work the team did, relationships that were built, and new skills that were developed. As part of closure, make sure to note any new skills and capabilities gained from their contributions.

LESSONS LEARNED REPORT

Without learning, we are doomed to repeat mistakes and forget success. Learning is a core function of project closing. Transferring learning from one project to the next will continuously increase the probability of success.

For many project managers, when they are given a new project, the first thing they reach for is a Lessons Learned Report from a previous project. The Lessons Learned Report along with the risk register from previous similar projects are some of the best assets a project manager can access to kick a new project off on the right foot.

Lessons learned are compiled throughout the project or at specific intervals, such as at the end of a life cycle phase. These are recorded in a lessons learned register. The intention is to improve project performance during the project rather than waiting until the very end to reflect on what is working and what isn't. Collecting lessons learned during the project sets aside time for the team to reflect on how to improve the project processes and the project performance.

The Lessons Learned Report is a compilation of the lessons learned during the project. It is organized and structured so that the organization and other project teams can use the information to perform better. The three standard questions used when gathering feedback about each topic are:

1. What went right that should be considered best practice and used during every project?
2. What went wrong and should be used as a leading indicator metric of project problems?
3. What should be done differently on the next project?

There are a number of different ways in which these questions can be worded depending on the topic, but, in general, these are the three primary questions used that are summarized for each topic. This information can be gathered from face-to-face interviews, email or web-based surveys, and from the review of the lessons learned documentation that is gathered throughout the project's life cycle.

Developing the Lessons Learned Report

The Lessons Learned Report should include an analysis of each aspect of the project performance, such as requirements, scope, cost, quality, resources, and so forth.

Requirements

Information on requirements includes requirements definition and management. You can evaluate how effective your requirements elicitation activities were. For example, were the elicitation techniques appropriate? were there missed requirements? Perhaps you used a new method for eliciting requirements that worked particularly well.

You can also evaluate the quality of requirements documentation. Were they clear and unambiguous, or were there examples of misunderstandings or miscommunication? Consider how well you did in managing your requirements. Was there requirements creep? Did any changes in requirements go through the Change Control Board? And lastly, did the final outputs and outcomes meet the criteria identified for acceptance?

Scope

Evaluate how well you were able to define and decompose the project and product scope. Decomposing product scope is often clearer than project scope because you are often decomposing a physical object into its component parts. Project scope includes all the work to complete the product scope. For example, did you correctly identify the work necessary for documentation, testing, risk management, and stakeholder engagement?

Many project managers are employing hybrid project management. This means that some deliverables are developed using evolutionary techniques and others are developed using predictive techniques. Lessons learned about how well the development approach worked with the deliverables is very useful.

Another aspect of scope management is ensuring that all the scope is delivered, and only the scope that is requested is delivered. This means that any additions or modifications, or even deletions go through the Change Control Board.

Schedule

Assessing the work for the schedule can include determining if your scheduling approach was appropriate for the project. For example, for project components that were evolving it would be appropriate to use a task board or other Agile concepts rather than documenting that information in a predictive or waterfall type schedule.

You'll also want to evaluate whether the scheduled tasks were decomposed to the appropriate level, if your schedule was fully resourced, if the dependencies were appropriate, and if the schedule was evaluated for efficacy. A schedule for a project that involves new technology that doesn't have sufficient reserve is not a viable schedule. On the other hand, maybe you found a more effective way to get the work done and you were able to use your schedule to level your resources and spread the work out so there weren't a lot of convergence and divergence points.

When controlling the schedule did you manage both the critical path and the near-critical paths? Did you utilize schedule reserve appropriately? Make sure to document any areas where milestones were missed and describe what caused the variance and how performance can be improved in the future.

Cost

Cost estimating can be a very challenging activity. When compiling lessons learned you should determine if you used the correct estimating technique for the various work components and if the estimates were reliable.

Another aspect to consider is whether the cost reserve was appropriate for the project. Think about whether you allocated reserve properly, such as to accommodate unplanned in scope work, or whether you used it to make up for a cost variance.

Cost control for smaller projects should consider how cost variance was addressed and the degree of cost variance. Performance within ±5% is usually considered very good performance. If you were on a larger project perhaps you used earned value management. If this was the case, you should evaluate how well it worked. Was the cost performance baseline appropriate? Were the methods to determine the earned value appropriate? Was the estimate to complete reliable or was it too optimistic or too pessimistic?

Quality

Quality planning is particularly important when there are strict requirements and measures needed for a product, or when a project is done under contract. Evaluate whether the quality planning, such as determining how requirements will be verified and validated, was appropriate. Maybe you used a new quality management technique or a new quality control technique. Document how well it worked. Perhaps it is something that will be useful in the future or perhaps it didn't work and you should note that as well.

For quality management you evaluate whether the quality processes were carried out as expected. Did people follow the quality plan or did they take shortcuts? Did the quality plan lead to the expected results? Consider how well product and project defects were handled. Was there a root cause analysis done? Were corrections put in place? Was there an excessive amount of scrap and rework? All these aspects should be documented in the lessons learned register and then summarized in the Lessons Learned Report.

Resources

You'll want to evaluate how well the management of physical resources was handled such as estimating the amount needed, when they would be needed, whether they were on hand as needed, and so forth. Also consider whether the amount estimated was sufficient or if there were greater amounts of scrap than expected, leading to a resource shortage and a budget overrun.

For team resources, consider how well the team worked together, how well you supported the team, and if the degree of management was appropriate for the project. In other words, did you balance leadership and management skills effectively? Any practices used to develop the team and create a high-performing team that worked should definitely be shared throughout the organization.

Risk

Risk management is a big topic and if it is not well handled it can be uncomfortable addressing it in a Lessons Learned Report; however, it's very important. You can start by looking at how well the risk register was populated versus how many unexpected events occurred. Also consider how effective the planned responses were in addressing the identified risks. Did the actions you took to avoid risks end up eliminating the risk? Were the planned mitigations implemented in a timely manner and were they effective? Also consider whether the way you managed risks was effective. For example, did you address risks on a regular basis, or just fill out the risk register at the start of every phase and then forget about it?

If significant risk events occurred, they should be addressed in the Lessons Learned Report. A thorough analysis of how they occurred, why they weren't identified, and the response should be documented. You may also want to consider summarizing the risk register at the end of the project. Note which risks occurred and how effective the responses were. If you are tracking opportunities as well as threats this is a good time to emphasize the importance of opportunity management and how that benefited the project and the organization.

Vendor Management

If you used vendors on your project, you should take a look at how well the procurement process worked. Was the SOW clear and complete? Did the source selection criteria lead to a good vendor selection? What about vendor performance? Was there sufficient oversight to ensure that the vendor performed per the contract? Were changes handled through a formal change control process and were they reflected in the contract? Were there any claims against the vendor or against your organization?

The process of writing a lessons learned report requires that you take time to reflect and discuss problems and issues with key stakeholders. It is often in this discussion that you can really start to distill the lessons. This time allows you to articulate an organized presentation that can be shared with future projects.

Using the Lessons Learned Report

The Lessons Learned Report should be compiled in tandem with the Project Closure Report. Part of closing the project is compiling the lessons learned, and then making sure the report gives a fair and accurate description of both good and not so good outcomes.

The information in the previous section is good starting place; however, it should be tailored to meet your needs. You can add, delete, or combine topics as needed. For example, you may want a separate section on change management, or you may not need the section on vendor management.

If your project was taking a new approach, such as utilizing artificial intelligence for document review, or trying a hybrid approach for creating the deliverables, make sure to include a robust description of how well it worked.

The Lessons Learned Report is saved, along with the lessons learned register, in a repository. Repositories can be as simple as a lessons learned folder, they can be a searchable database, or anything in between. Since the Report is done at the end of the project, it is used for future projects, rather than the current project.

POSTMORTEM REVIEW

Sometimes called the Post Project Review or Post Implementation Review, the Project Postmortem goes by many names. Regardless of what you call it, the Project Postmortem is a review of a project after the project closure acceptance, after all project closure activities are complete, and after the project has been in operational mode for a period of time.

With the lessons learned from past projects helping to increase the probability of future project success, we can see the postmortem work serving as a linchpin in an organizational knowledge management process.

The opportunity for learning from the postmortem review process is significant. It is because of the learning aspect of postmortems that most organizations associate these processes as being part of knowledge management. The purpose of such a review is to determine the following:

1. Was the project successful?
2. Were all closure activities handled properly, especially any final hand-offs from the project team to the operations team or customer?
3. Were lessons learned captured and transferred to project teams across the enterprise properly?
4. Has the project achieved planned operational outcomes – the business goals and objectives outlined in the project's business case?

Figure 17.1 illustrates how the Project Closure Plan and Checklist precedes the Project Closure Report and Lessons Learned Reports. The Postmortem Review occurs after all other closure activities are complete.

Conducting the Postmortem Review

The three steps most associated with the Postmortem Review are: (1) gather feedback from project and operational teams, (2) organize and facilitate a meeting among the teams and key stakeholders, and (3) capture the meeting in the form of a Postmortem Report.

Gathering Information

Information from the Project Closure Report and the Lessons Learned Report provides a good foundation for the Postmortem. The project sponsor and operations leadership

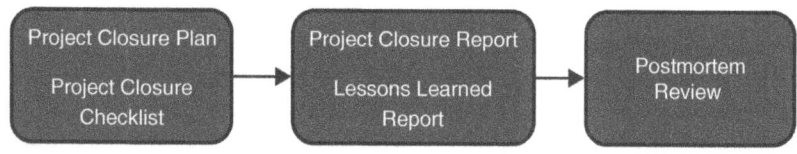

Figure 17.1: The Flow of Closure Work

should identify a well-respected and non-biased person to facilitate the postmortem meeting. We do not recommend that the project manager be the postmortem review facilitator.

Gathering feedback ahead of the meeting will help to expedite the meeting and provide the facilitator with a sense as to how the meeting will go—will it be emotionally charged and negative or will it be productive, future-focused, and positive.

Conducting the Meeting

The postmortem meeting can either feel like a celebration or a poor performance review. It will be a celebration if things have gone well. However, in the case of poor project execution, missing business goals while in operations, personality conflicts, organizational politics, or a number of other variables, the postmortem meeting could be negative. It is for this reason that a neutral and expert facilitator is recommended to oversee such meetings. This individual can summarize feedback gathered ahead of the meeting and dive into the details necessary to obtain real meaning and insight on what to do better during future projects. The aim of the meeting is to uncover ways to make project work better and strengthen the culture of the individuals involved. Expert facilitators can navigate the challenging conversations of an otherwise negative Post-mortem Review and have participants leave the meeting feeling good about improvements for future projects.

Documenting the Postmortem Report

The Postmortem report is an important artifact. Best practices suggest the document not be summarized in bullet points, but rather written in a narrative form of what to do and what not to do. Bullet points run the risk of being too ambiguous, whereas a narrative that is properly categorized can provide a richer context of lessons learned, thus making them easier to read, easier to contextualize, and easier to apply. Further, this report should be provided to project and operational teams beyond those represented in the postmortem meeting. The idea of the report is to benefit all teams as an enterprise asset, not just those involved in the postmortem.

Best Practices for Postmortem Success

1. *Use experts:* since it is often difficult to facilitate a meeting and document the meeting and do so without bias, use an expert facilitator, scribe, and others during the postmortem review.
2. *Ensure representation:* make sure the key members from the project team and operational team are present and comfortable in sharing their thoughts, ideas, and opinions openly and productively.
3. *Everyone contributes:* conduct the post mortem in a way that everyone participates without peer pressure or senior leader persuasion.

(Continued)

4. *Work from facts:* make sure facts are known and have any comments facilitated to the point of it being fact-based rather than subjective opinion.

5. *Focus on the future:* while the postmortem is a reflection of past events, the primary focus is on how to make future projects better and therefore the majority of time should be spent on what should be done differently next time.

6. *Detail the conversation in narrative form:* rather than high-level bullet point summaries of the conversation, have the meeting detailed in a narrative report because stories make for easier learning than bullet points.

7. *Broadcast your results:* be sure to share the postmortem report with other project teams and archive in an easily accessible database for other project managers to use.

Many facilitators of postmortem reviews find that using a checklist can be helpful throughout the postmortem process—from gathering feedback to conducting the meeting to documenting the final report. The sample checklist in Table 17.7 offers questions are more far reaching than the Lessons Learned questions noted earlier in this chapter.

Depending on the response to each question in the postmortem checklist questionnaire, follow-up or probing questions can (and should) be asked. For example, if any question is answered "no" the facilitator should ask, what should be done differently on the next project to ensure this occurs? The facilitator could also ask if there were any leading indicators of a problem.

Table 17.7: Project Postmortem Checklist Questionnaire

Category	Question
Project planning	1. Did the business case clearly detail the problem or opportunity? 2. Were business goals and project objectives clear and measurable? 3. Was the project scope, schedule, budget, and quality clearly detailed? 4. Were project plans detailed, accurate, and usable?
Customer focus	5. Was the voice of the customer evident in all phases of the project? 6. Did deliverables and milestones meet customer (and stakeholder) target expectations? 7. Was communication with customers effective?
Deliverables	8. Did the outcome of the project meet your expectations? 9. What gap or variance (if any) exists between the expectations and the final deliverable? 10. Were project monitoring and managing efforts effective?
Communication	11. Was there an effective communication plan for the project? 12. Was there proper stakeholder management (analysis and monitoring)? 13. Was communication (to the team, to sponsors, to all stakeholders) effective?
Decisions	14. Was there a clear decision-making process (including an escalation process) in place and used? 15. Were decisions made fast enough?

Using the Postmortem Review

The actual postmortem work effort is conducted three to six months after the final handoff from the project team to the operations team or customer. The timing between the project handoff and postmortem work is necessary in order to allow the operations team enough time to realize the value and benefit from the project.

While the tools and templates outlined for postmortem work facilitate a reflection of the past, the focus during this work is on the future. The reflection of the past can mostly be gleaned from the Project Closure Report and the Lessons Learned Report. Therefore, the effort and workload associated with postmortem activities should be minimal. While the workload is often minimal, the results from the work can be very dramatic. The tools, and especially the Postmortem Review meeting, should focus on leveraging experiential knowledge to make future projects better (more efficient and effective), create higher performing project teams, and increase business value.

Most postmortem meetings have a duration of anywhere from one to four hours. The larger and more complex the project, the longer the postmortem meeting may require. Although it may be assumed at these meetings, it is important for the facilitator to emphasize the need for an honest, candid, and objective discussion and to focus on process and not people. These behavioral ground rules serve as guidelines for participants and the facilitator. Such ground rules establish a constructive atmosphere for discussion and learning. These ground rules also help to prevent personal attacks that disengage participants and disable learning.

In addition to the information gathering ahead of the meeting and ground rules for use during the meeting, a prerequisite for an effective Postmortem Review is a well-crafted agenda. The agenda will vary based on findings from the information gathering work effort. The following bullet points, however, offer a sample agenda for a postmortem review.

- Welcome and introduction of everyone
- Review of the ground rules by the facilitator
- Summary of information gathering findings from the facilitator (note, this information should be sent to participants in advance of the meeting if at all possible)
- Review and rank issues and critical success factors
- Create a "what went wrong" list
- Create a "what went right" list
- Detail opportunities for improvement with specific (actionable) recommendations
- Outline specific points of communications and next steps.

The facilitator can manage this agenda in a number of ways. For example, all items can be addressed in an open, round-table conversation. As an alternative, the facilitator could conduct this meeting as a workshop in which case participants are much more active. However the postmortem review is conducted, it should fit the organizational culture and conversational tone necessary to achieve the end results, which is individual, team, and organizational learning.

INDEX

Project Management Toolbox: Tools and Techniques for the Practicing Project Manager,
Third Edition. Cynthia Snyder Dionisio, and Russ J. Martinelli.
© 2025 John Wiley & Sons, Inc. Published 2025 by John Wiley & Sons, Inc.